SpringerBriefs in Applied Sciences and Technology

More information about this series at http://www.springer.com/series/8884

Gonçalo Dias · Micael S. Couceiro

The Science of Golf Putting

A Complete Guide for Researchers,
Players and Coaches

Gonçalo Dias
Faculty of Sport Sciences and Physical
   Education
University of Coimbra
Coimbra
Portugal

Micael S. Couceiro
Ingeniarius, Lda., Artificial Perception for
   Intelligent Systems and Robotics
   (AP4ISR)
Institute of Systems and Robotics (ISR),
   University of Coimbra
Mealhada
Portugal

ISSN 2191-530X   ISSN 2191-5318   (electronic)
SpringerBriefs in Applied Sciences and Technology
ISBN 978-3-319-14879-3   ISBN 978-3-319-14880-9   (eBook)
DOI 10.1007/978-3-319-14880-9

Library of Congress Control Number: 2014959824

Springer Cham Heidelberg New York Dordrecht London

Printed on acid-free paper

Springer International Publishing AG Switzerland is part of Springer Science+Business Media
(www.springer.com)

Foreword

Golf is one of the most popular sports worldwide. For example, in 2013, the Professional Golfers' Association (PGA) estimated that there were over 26 million golfers solely registered to play in the United States of America (USA). This book is a necessary addition to the scientific and technological literature on one of the most important actions in golf: the 'putt'. Putting requires deftness, accuracy, and force control, and utilizes a cyclical relationship between perception (visual, proprioceptive, haptic, and acoustic) and action since it is the most precise aspect of the sport of golf. In fact, the PGA considers it as the most vital skill to master in the sport, representing about 43 % of the total number of strokes in a competitive round of golf. However, despite golf's ever-increasing popularity worldwide, the role that science and technology can play in enhancing performance and acquiring skill in putting needs to be better understood.

This book by Gonçalo Dias and Micael Couceiro provides a scientific and technical treatise on the process of golf putting, emphasizing the different pathways to achieving successful performance outcomes in this task. The engineering perspective for studying the *process of successful coordination* in golf putting is novel and warranted. There are many coach and player-oriented performance manuals available in the practical literature, especially on the golf swing, which fail to provide a bioengineering perspective addressing the human-equipment interface. This Springer Brief on the topic of golf putting is a useful contribution to the applied science literature since it focuses on a most important part of the game identified by performance analysts. This specialized Brief will be of major interest for engineers, sport scientists, human movement scientists, and many coaches/players who are interested in developing putting skills. The focus on measuring performance of the golf putting action through use of product and process variables is an excellent novel idea adopted in this book, which complements the existing literature on human movement science and sport science from a dynamical systems orientation, complexity sciences viewpoint, and an ecological dynamics perspective. It is my pleasure to write this preface since it is apparent that Gonçalo Dias and Micael Couceiro adopt a relevant and much-needed combination of engineering and sports science expertise, which can contribute to the bioengineering perspective proposed. This book benefits from the academic perspective that the authors adopt, as well as their willingness to discuss the relevant research and applied sport science literature on this topic. The book utilizes their published research on process and product variables in analyzing sport performance and human movement. It is my belief that the material discussed herein will enhance the performance, practice and training of golfers at elite and sub-elite levels.

Sheffield, UK, December 2014   Keith Davids

Acknowledgments

The authors would like to thank Prof. Rui Mendes, Prof. João Barreiros, Prof. Keith Davids, Prof. Guilherme Lage, Prof. Hugo Espírito Santo, Prof. Nuno Barreto, Eng. André Araújo, Eng. Samuel Pereira and Mr. António Dias for their technical support.

Technical Note

Undeniably, this publication fills an empty space in the bookshelf of all professional golfers. As the authors claim, the book presents both educational and training perspectives as well as professional and scientific aspects. The writing is accessible and the diagrams are splendid and deliberately prepared for the purpose of this manual. The book gives the impression that for 'ordinary men' the secret is far beyond hitting the ball with a club and hoping it enters the hole, which, in fact, is quite true.

Contrary to what one might think, putting is a rather complex motor skill, which involves a set of motor performance variables and measures that should be studied to understand the reason, or reasons, why one misses the hole. In spite of this, it is important for the coach and the athlete to know how it is possible to 'calibrate' and 'tune' their putting performance in the context of motor learning, training and competition. This remarkable compendium dissects all this information around putting in a scientifically clear and sound manner. The degree of innovative metrics that the authors have published over the past years in relevant journals, such as the *Journal of Motor Control* and the *Journal of Motor Behavior*, as well as in many chapters of Springer books, are noteworthy. This 'ecological perspective' of golf putting is described throughout this book, in harmony with the constraints arising from the action of the practitioner, the task and the environmental conditions. Additionally, the authors briefly discuss the different practice conditions, the intra- and inter-individual variability emerging from this pendulum-like motion, the influence of the contextual interference phenomenon, and the motor skill organization of experts and novices. All these topics end with some practical implications associated with learning and motor control.

The book also describes new technological outputs, highlighting the role of cameras, inertial sensors and other tools that are able to provide golf putting analysis as not seen before. In this particular aspect, one cannot neglect the outstanding contribution of the authors in the development of the new device *InPutter*. This engineered golf putter, invented by *Ingeniarius, Lda.*, revolutionizes the way that putting may be analyzed.

In summary, this book fills an emerging gap around the learning and training of golf putting portrayed in the literature. It is a remarkable contribution for researchers, coaches and athletes who wish to better understand the execution of putting in the laboratory and in real-life contexts of teaching and learning.

Finally, it is my belief that this book can effectively decode the 'science behind golf putting'.

Hugo Espirito Santo
European Champion of Pitch and Putt

Contents

About the Authors

Gonçalo Dias obtained both M.Sc. and Ph.D. degrees on Sport Sciences at the Faculty of Sport Sciences and Physical Education of University of Coimbra and the B.Sc. degree at the Coimbra College of Education from the Coimbra Polytechnic Institute. He is currently a Professor of Advanced Topics in Motor Learning at the Faculty of Sport Sciences and Physical Education of University of Coimbra (Master Course in Youth Sports Training) and a member of the Scientific Committee within the same institution. Over the past 6 years, he has been conducting scientific research on several areas associated to sports and health, namely physical activity among older adults, motor control, sport sciences, and others. Besides research and lecturing, he has been in charge of the organization of events, both in the public and private domains, as coordinator of the County Sports Office from Arganil, Portugal. Additionally, he is responsible for the program "Physical activity to older adults," which brings together more than 100 elderly aged between 65 and 100 years old. This activity resulted in the publication of two books and several articles that portray the work of his research in 20 private institutions of social solidarity in Portugal, being currently considered one of Portugal's leading experts in this field.

e-mail address: goncalodias@fcdef.uc.pt

Micael S. Couceiro obtained the B.Sc., Teaching Licensure, and Master degrees on Electrical Engineering (Automation and Communications), at the Coimbra School of Engineering (ISEC), Coimbra Polytechnic Institute (IPC). He obtained the Ph.D. degree on Electrical and Computer Engineering (Automation and Robotics) at the Faculty of Sciences and Technology of University of Coimbra (FCTUC), under a Ph.D. Grant from the Portuguese Foundation for Science and Technology. Over the past 6 years, he has been conducting scientific research on several areas, namely robotics, computer vision, sports engineering, and others, all at the Institute of Systems and Robotics (ISR-FCTUC) and RoboCorp (IPC). This resulted on more than 25 scientific articles in international impact factor journals and more than 50 scientific articles at international conferences. Besides being currently a researcher at ISR, he has been invited for lecturing, tutoring, and organization of events (e.g., professional courses, national and international conferences, among others), both in the public and private domains. He is a co-founder and currently the CEO of Ingeniarius—a company devoted to the development of smart devices, from which performance analysis in sports has been one of its main markets.

e-mail address: micael@ingeniarius.pt

Chapter 1
Introduction: Pendulum-Like Motion in Sports

Abstract This starting chapter aims to introduce the main mechanics behind *pendulum-like motion*, by bridging the gap between theory and its applicability in sports. To that end, we start by presenting a brief historical background of this well-known physics case study steered by the legacy of *Galileo Galilei*, thus showing how this phenomenon was innovative for the evolution of science in several research areas. This is followed by a description of some sport modalities that intrinsically benefit from a pendulum-like motion, paying particular attention to golf putting.

Keywords Golf putting · Pendulum-like motion · Learning · Constraints · Performance

1.1 Characteristics of Pendulum-Like Motion

When we talk about pendulums it is almost mandatory to go back to their origin by observing the amazing work of *Galileo Galilei* (Fig. 1.1). This scientist decisively contributed to the deepening of several research areas, such as physics, astronomy and mathematics (Baker and Blackburn 2005).

The motion of suspended bodies started to appeal to Galileo's curiosity in his early life. The story goes that his interest arose in 1588 when he was attending the church at the Cathedral of Pisa. At that time, Galileo was surprised to observe that the chandeliers hanging in the Cathedral would take the same time to complete an oscillation period regardless of their angular displacement. Several years later, in 1602, Galileo conceived, for the first time, the idea of the *isochronism of the pendulum*, in which he argued that the period of oscillation of a pendulum is independent of its angular amplitude (i.e., *angle of swing*). In addition, Galileo also concluded that the pendulum would return almost to the same angular displacement from which it had been dropped; this is nowadays accepted as a manifestation of

© The Author(s) 2015
G. Dias and M.S. Couceiro, *The Science of Golf Putting*,
SpringerBriefs in Applied Sciences and Technology,
DOI 10.1007/978-3-319-14880-9_1

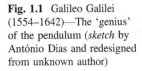

Fig. 1.1 Galileo Galilei (1554–1642)—The 'genius' of the pendulum (*sketch* by António Dias and redesigned from unknown author)

energy conservation—a concept not yet introduced at that time (Baker and Blackburn 2005).

When carrying out further experiments on pendulums, Galileo also noted that the duration of the pendulum oscillation does not depend on the weight of the body attached to the end of the wire. Knowing that the pendulum motion is caused only by gravity, Galileo confirmed this effect by freely dropping two stones of different weights from the Tower of Pisa, and observing that they would both take the same time to reach the ground. These findings were controversial at the time, as they contradicted the conclusions and fundamentals of Aristotle (Baker and Blackburn 2005).

Interestingly, Galileo made measurements of the pendulum motion period using his own heartbeat as a reference. In 1641, when Galileo was practically blind, it occurred to him that it should be possible to adapt the pendulum theory to clocks, using weights or springs. He believed that the flaws of conventional watches at that time could be corrected by the periodic motion intrinsic to pendulums. Indeed, it was that discovery that fostered the design of more accurate clocks, since the period T of the simple pendulum depends only on its length L and the angular displacement θ_0 as:

$$T \approx 2\pi\sqrt{\frac{L}{g}} \cdot \sum_{n=0}^{\infty}\left[\left(\frac{(2n)!}{(2^n n!)^2}\right)^2 \cdot \sin^{2n}\left(\frac{\theta_0}{2}\right)\right], \tag{1.1}$$

where g is the acceleration of gravity (typically ~ 9.8 ms^{-2}). Note that this is just one of the many mathematical models of the pendulum period, known as the

Legendre polynomial solution. For more details about possible alternatives, please refer to Baker and Blackburn (2005). The obvious advantage of Eq. (1.1) is that it only depends on two variables that are easy to control—the length L and the angular displacement θ_0 of the pendulum.

These insights obtained by Galileo were followed by many other researchers and applied to many different fields, including several sport modalities.

1.2 Sports

The literature is scarce about the operational description of pendulum mechanics adapted to the sports context. In fact, there is a huge void around the topic which reinforces the need to deepen this concept in this book.

Due to the nature of this book, we will mainly focus on individual sports, e.g., bowling, pétanque and, as expected in this book, golf; these show a higher potential in terms of pendulum-like motion than that we find in collective team sports (e.g., soccer).

1.2.1 Bowling

Bowling is a well-known sport throughout the world. The main objective is to throw a ball on a track and try to hit 10 pins placed in a triangular formation. The control of this pendulum-like motion is made by the action of the dominant arm (Fig. 1.2).

Although it seems easy to perform a bowling movement, it is necessary to learn to focus mainly on obtaining a pendulum-like motion with the arm that can be effective during competition. On the other hand, this may only be ensured by maintaining a constant repetition of numerous training trials aimed at the mechanics of the movement.

1.2.2 Pétanque

In the 'short' game of pétanque, the player is meant to throw a metal ball on a sand or grass surface, and get as close as possible to a small wooden ball (denoted the *cochonette*) located on the ground (Fig. 1.3). Similar to bowling, the control of this pendulum-like motion is achieved through the action of the dominant arm. However, while in bowling, the thrower holds the ball underneath the object due to its weight and dimensions, in pétanque the hand is over the ball. This aspect influences how the player performs the pendulum motion while executing the action.

Fig. 1.2 Pendulum-like motion: bowling example (*sketch* by António Dias and redesigned from unknown author)

Fig. 1.3 Pendulum-like motion: pétanque example (*sketch* by António Dias and redesigned from unknown author)

This traditional French game requires a great level of coordination and visuomotor action, since the pendulum motion, by simply gravity, needs to toss the ball to a desired position. In this situation, expert athletes can adapt the toss force by adjusting the angular displacement of their arm, thus adapting to the demands of the game.

1.2.3 Golf Putting

No other sport portrays pendulum mechanics like *golf putting*. Without doubt, if we were to elect a motor skill that comes closer to a perfect pendulum-like motion in its motor performance, then putting would be the winner. This evidence is clearly present in the work of Pelz (2000), which shows exhaustively how pendulum mechanics influences all phases of this movement; its properties change based on the morphological, biomechanical and functional features that substantially vary from player to player in terms of range of motion. Due to its relevance, a more extensive study on the properties of pendulum mechanics and a brief description of the current state of the art on the *science of putting* are to be found in the next chapter.

1.3 Practical Implications

Analysis of the pendulum mechanics of putting has practical implications for both coaches and players, as it allows a better understanding of the execution of motor skills. Hence, understanding the physics behind the pendulum can contribute to adequate classification of all the movement phases inherent in putting and its associated variables. This type of pendulum action is common to other sports that use the upper limbs to achieve success.

References

Baker GL, Blackburn JA (2005) The pendulum: a case study in physics. Oxford University Press, Oxford

Pelz D (2000) Putting Bible: the complete guide to mastering the green. Publication Doubleday, New York

Chapter 2
Golf Putting

Abstract This chapter introduces golf putting by describing its movement phases and all other relevant details. As such, we describe how we can learn this skill of driving in different conditions of practice variability, thus showing the 'science' behind this action. To do this, we will use the area of motor control and explain how players can optimize their learning performance and address the constraints that emerge from the task and environment.

Keywords Golf putting · Variability · Constraints · Performance

2.1 Description of the Movement

Golf is one of the most well-known short games in sport, where competing players need to introduce the ball into the hole with the fewest number of strokes. According to Pelz, the putting technique, or simply putting, is defined as a light golf stroke made on the green in an effort to place the ball into the hole (Pelz 2000). Note that this movement represents approximately 43 % of the strokes in a golf game (Alexander and Kern 2005).

Authors such as Pelz (2000), Hume et al. (2005), Couceiro et al. (2013) and Dias et al. (2013) divide the execution of the golf putting movement into four phases (Fig. 2.1):

(1) *Backswing* movement of the putter upwards and backwards in relation to the ball. This phase is necessary to position and align the golfer's hub center and the club head.
(2) *Downswing* movement of the putter downwards and forwards in relation to the ball. This phase starts where the backswing phase ends, and finishes immediately before the club head strikes the ball in the correct plane under maximum velocity.
(3) *Contact/ball impact* time instant in which the club head strikes the ball.

© The Author(s) 2015
G. Dias and M.S. Couceiro, *The Science of Golf Putting*,
SpringerBriefs in Applied Sciences and Technology,
DOI 10.1007/978-3-319-14880-9_2

Fig. 2.1 Phases of golf putting: **a** backswing; **b** downswing; **c** ball impact; and **d** follow-through. Sequence executed by the European Champion of *pitch* and *putting*: Hugo Espirito Santo

(4) *Follow-through* starts immediately after the ball impact and consists of the deceleration phase by benefiting from eccentric muscular actions, i.e., an inertial phenomenon inherent in golf putting (Hume et al. 2005).

Pelz (2000) shows that this pendulum movement is described by a simplistic and yet efficient model, thus being easy to learn when compared to ballistic actions, such as the tennis serve (Mendes et al. 2012). The configuration behind putting has been commonly described by a triangle, formed by an imaginary line that connects the shoulders of the golfer with his/her hands: 'The Putting Bible'.

Supported by *Newton's Universal Law of Gravitation*, Pelz (2000) proves that the force of gravity seems to create, on its own, an adequate optimal velocity for the execution of putting, without any incremental velocity or force applied by the golfer.

By means of this pendulum technique, golfers can adopt a rhythm similar to the cadence of a pendulum clock or a metronome,[1] maintaining a constant velocity during the process of motor execution. However, it should be noted that the golf putting performance substantially varies from player to player, depending on their morphological and functional features (Schöllhorn et al. 2008; Karlsen et al. 2008). For instance, as stated in Chap. 1, the length of the rod, which in this case is related to the length of the golfer's arms and club, influences the pendulum's maximum velocity. Besides this morphological feature, the golfer's reasoning and cognition also affect the end result. For example, the golfer needs to adapt the initial angular displacement (initial position of the downswing) based on the distance to the hole. As a rule of thumb, the larger the angular displacement, the larger the maximum velocity when the putter hits the ball (see Chap. 1).

Although Pelz (2000) states that golf putting is a simplistic movement with little or no struggle regarding its motor execution, he also emphasizes that within such deceptive simplicity, one needs to take into account a wide range of variables that can make this quite a complex movement.

Confused with this description? Well, let us then explain to what extent the simplicity of golf putting can be actually quite complex!

2.2 Learn How to Putt

This section describes the learning process behind golf putting, covering its basis by following the work of Pelz. The authors decided upon this choice since this researcher made a remarkable contribution within the laboratory context. Please refer to 'The Putting Bible' for a deeper understanding of the teaching and learning of golf putting (Pelz 2000).

2.2.1 The 'Science' Behind the Learning

Golf putting can be learned in many different ways and with many different methods. However, the individual *variability* underlying human motor behavior makes this

---

[1] According to Merriam-Webster dictionary, a metronome is 'a device that makes a regular, repeated sound to show a musician how fast a piece of music should be played'.

movement different from player to player, since the morphological and functional characteristics are distinct. Therefore, to better understand the concept of biological and functional variability, let us introduce some preliminaries around this subject.

Variability is a key feature of any biological system and plays a key role in human motor behavior (Newell and Corcos 1993). When we talk about the variability that underlies the human body, we have to realize that the interaction established between the various bones comprising our skeleton results in 244 degrees of freedom (DOF)— the movement of a human arm requires 7 DOF and, if analyzed in terms of muscles, the number of DOF can segment into up to 26 possible combinations (Rose 1997; Magill 2011).

With regard to golf putting performance, the variability may be associated with how the task or the golfer is 'disturbed'. For example, if we change the mass of the putter head, we may increase the 'noise' in the putting performance, which, by itself, is already quite complex to analyze. Under those conditions, we can identify two types of human variability—*intra-individual variability* and *inter-individual variability*. Intra-individual variability is related to the morphological, physiological or psychological features (profiles) that depict a particular golfer. The inter-individual variability depicts performance trends among several golfers during the execution of the same movement under the same conditions (Dias et al. 2011).

With this in mind, Pelz (2000) states that there are several strategies to learn how to putt under several conditions of variability, including:

1. hitting the ball at different distances to the hole, contemplating various paths (linear or curvilinear) and slopes (ascending or descending);
2. executing the putting under different atmospheric conditions (e.g., sun, rain, wind, snow and cold);
3. training to putt under different green and practice conditions (green with short grass, long grass or mishandled grass, with holes, squashed by the spikes of golf shoes, with sand, etc.);
4. executing putting with different types of balls ('soft' and 'stiff'), using clubs of different lengths (small, average, large).

Given these arguments, the reader should be wondering if all this variability does not work as a 'negative constraint' for the golfer during the execution of golf putting. However, as we will see through the model of Karl Newell, by 'constraining' the action of the athlete we may positively influence his/her action system, thus providing new mechanisms of motor adaptation and self-regulation in the course of his/her learning.

2.2.2 'Constrain' to Learn

Newell's model (1986) shows that the constraints contribute to the regulation of human movement dynamics. These constraints are related to the connectivity between each component, establishing a particular form of self-organization. As

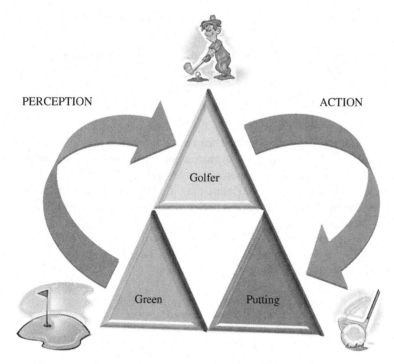

PERCEPTION   ACTION

Golfer

Green   Putting

Fig. 2.2 Constraint model—adapted from Newell (1986)

such, they should not be seen as a negative influence on behavior, but rather as a set of limitations that may influence the action system (Davids et al. 2008). After validating that the constraints influence the dynamic response of the human movement, Newell (1986) introduced the *constraints-based approach* by classifying them into three different categories—the athlete, the environment, and the task. This model shows that both learning and motor performance emerge from the interaction that occurs between these three categories (Fig. 2.2). Therefore, the dynamics established between the constraints is a factor that needs to be taken into account during the performance of this movement.

By bridging a relationship between Newell's constraints and the learning of golf putting, one can easily argue that both the morphological (height, weight and height) and functional (motivation, fatigue, etc.) characteristics of the players may influence the angular displacement applied to the putting and, as a consequence, to the force, acceleration and velocity of the movement throughout the putting performance. These aspects relate to the characteristics and profiles that distinguish each individual practitioner in the motor execution process. On the other hand, the

Table 2.1 Newell's constraint model (1986) adapted to the putting performance

Player	Environment	Task
Morphological characteristics: height, weight and height	Weather: wind, rain and morning dew	Irregularities of the green: slopes and texture of the grass
Functional characteristics: motivation, fatigue, among others		

irregularities of the green (slopes and texture of the grass) may also compel the golfer to adjust his/her technique to overcome the restrictions imposed on the task, thus inevitably influencing the putting performance (Table 2.1). Finally, the atmospheric conditions (wind and rain) may also constrain the displacement of the ball on the green, thus forcing the player to adapt to different situations during the game (Pelz 2000).

Within the three types of constraints presented by Newell (1986), the ones that shows a higher degree of manipulability and that have been more studied in the literature are the ones provided by the task.

2.2.3 Practice Tips

In line with the previous sections, this section summarizes some practice tips so as to foster the learning of golf putting. Let us start by the perception of the task. Pelz (2000) highlights the need to 'read' the green and confirm its main characteristics (e.g., slopes). This 'routine' is very important in the decision-making of the player and can influence the outcome. It is often common to watch expert players on the PGA Tour kneeling on the green and carefully visualize the state of the grass. All these factors have a decisive influence on how the golfer will hit the ball and adjust its action, i.e., the angular displacement of the putter and all other related process variables of motor execution. Naturally, this will also have repercussions on the final location of the ball (see the concept of the radial error described in the next section).

As well as the green, the shape and size of the putter is also worthy of appreciation since it should 'fit' the athlete's morphological features, namely the size of his arms and hands. A good grip is the one that best suits both the morphological and functional characteristics of the golfer, offering the highest probability of succeeding in the task (Pelz 2000). In that sense, from an instrumental point-of-view, one can verify that the putter plays a crucial role in the motor performance of this action (Karlsen et al. 2008; Mackenzie and Evans 2010). Hence, researchers such as Gwyn et al. (1996) state that the range of putters available on the market is quite diversified and broad, and one can find almost all possible sizes and materials (e.g., titanium, graphite, steel). This aspect is different from the choice of golf balls, whereby a different material may result in a completely different outcome.

Karlsen et al. (2008) and Karlsen (2010) state that the market trend over the last decade depicts an increasing number of putter designs. Their survey shows that some researchers compared putting performance using different putter designs. However, Gwyn et al. (1996) found no statistically significant differences in the putting accuracy when comparing a traditional blade-putter and a cylindrically formed putter head. Karlsen and smith (2007), Karlsen et al. (2008) and Karlsen (2010) devoted their studies to understanding the influence of putter length on performance. With this in mind, Pelz (1990) also compared putting performance using long and conventional putters, confirming that a long putter was best on short putts (0.9 m), equally good on medium putts (2.7 m) and worse on long putts (6.1 m). However, Gwyn and Patch (1993) did not find any significant differences regarding the length of the putter. Alternatively, Brouliette and Valade (2008) compared grooved and milled putter faces with different loft for skid distance; however, skid distance did not seem to be reduced by grooves, suggesting that the design of the putter head can also affect putting performance (cf. Karlsen et al. 2008; Karlsen 2010). Nilsson and Karlsen (2006) found that a specially designed wing-shaped putter performed better than a blade and a mallet putter, both in distance and direction on horizontal miss hits.

In spite of this, Pelz (2000), Karlsen et al. (2008) and Mackenzie and Evans (2010) showed that golf club manufacturers are starting to exploit the possibility of having 'adaptive' putters. However, little information is provided on how this may be accomplished, especially considering the variable weight and height of golfers, and how they may achieve an 'optimum putter' using the same putter. This evidence allows us to conclude that it is necessary to deepen the research around putter design and further assess its influence on putting performance.

Besides the perception of the task and the putter, the execution of motor skills needs some special care per se. Motor stability is of crucial importance, in such a way that the golfer can evolve while learning putting.

As such, it is once again recommended to follow the physics of a simple pendulum governed by Newton's Universal Law of Gravitation. One way to promote this is to film the motor performance and to analyze it later to identify and correct any technical inconsistencies.

Given the evidence presented above, the *5-Step Routine* by Pelz (2000) is strongly recommended to foster the learning of golf putting (Fig. 2.3).

After the basis of how to execute golf putting, now comes the time to understand how we may evaluate it. Note that such insights presented in the next section may be used not only in the research perspective, but also for training purposes.

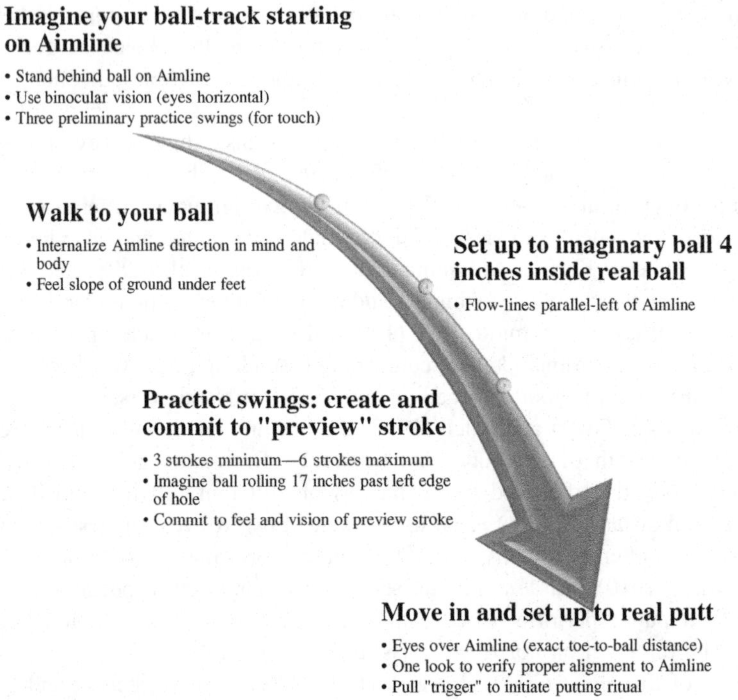

Imagine your ball-track starting on Aimline

• Stand behind ball on Aimline
• Use binocular vision (eyes horizontal)
• Three preliminary practice swings (for touch)

Walk to your ball

• Internalize Aimline direction in mind and body
• Feel slope of ground under feet

Set up to imaginary ball 4 inches inside real ball

• Flow-lines parallel-left of Aimline

Practice swings: create and commit to "preview" stroke

• 3 strokes minimum—6 strokes maximum
• Imagine ball rolling 17 inches past left edge of hole
• Commit to feel and vision of preview stroke

Move in and set up to real putt

• Eyes over Aimline (exact toe-to-ball distance)
• One look to verify proper alignment to Aimline
• Pull "trigger" to initiate putting ritual

Fig. 2.3 Five-step routine—adapted from Pelz (2000)

2.3 Evaluating Golf Putting Performance

State-of-the-art research shows that golf putting is focused on performance analysis, thus confirming its relevance towards evaluating a complex motor ability which differs from player to player. In that sense, many authors such as Beilock and Carr (2001), Singer (2002), Beilock and Gonso (2008) etc., indicate that the best golfers in the world train their psychological skills and new strategies of mental visualization (i.e., mental imagery). This kind of reasoning allows control of anxiety and pressure, which are crucial during competitions. Pelz (2000) shows that expert golfers can be distinguished from inexperienced golfers because, among other conditions, they are able to successfully follow these strategies:

(1) control both pressure and anxiety states with greater ease, often using imagery and mental visualization of the action and the outcome beforehand;
(2) obtain high levels of concentration and intrinsic motivation, adopting behavioral routines during the task;
(3) perform golf putting in an autonomous and involuntary way, without allocating too much time thinking of its execution.

Golf putting also encompasses other relevant variables within the performance context, such as stability, routine,[2] attitude and rhythm (Pelz 2000), as well as other aspects of personality, learning ability and motivation in the execution of this movement (Jonassen and Grabowski 1993; Orliaguet and Coello 1998; Vickers 2004).

When we analyze golf putting, it is important to consider the largest amount of variables that may influence its performance. Therefore, it is common in the field of motor control to consider both product and process variables to accurately describe a particular problem related to the learning or motor coordination of players. Product variables are directly related to the performance measure of the final product or outcome (distance between the final position of the ball to the hole). On the other hand, process variables are related with the implementation and operation of the motor execution (acceleration of the putter when hitting the ball). Most of the research around golf putting proposes or re-evaluates metrics within these two domains. Chapters 4 and 5 will describe most of the metrics proposed to date in the evaluation of both product and process variables.

2.4 Practical Implications

In our conceptual framework, we assume that research around putting is rather scarce from a perspective that constraints emerging from learning and training may be beneficial to a golfer's performance. In that sense, this chapter gives the first steps to help understand the most adequate way to analyze putting in a laboratory and real-life context. Although golfers and coaches are aware that there is a number of 'ecological' variables that are inevitably present on the green, it is not known how to take them into consideration in order to win. Furthermore, we know that process variables inherent in motor execution, and how these influence putting, can be decisive in the final outcome, i.e., what we describe as 'product'. Thus, in terms of practical implications, there is little research that portrays how a certain player (novice/expert) can, or cannot, properly 'read' the constraints of the green and clearly perceive the trajectory that the ball may take over the hole. Yet, this can be the difference between winning or losing a game at the highest level!

The above-mentioned considerations are still practical implications for coaches and players, as they allow better understanding of how a seemingly simple pendulum action, as in the case of putting, can be rather complex and include a large set of variables of a physical and cognitive nature. We propose that putting learning and training are enhanced depending on the characteristics of the players and their

---

[2] The routine in golf can be characterized by a sequence of programed actions to prepare for the performance of the golf putt. For example, there are golfers who walk the green and 'read' it from several places; they simulate the performance of the putt without hitting the ball (putting preparation); they examine the hole from several viewpoints; they close their eyes to mentally visualize the movement and imagine the trajectory of the ball, etc.

morphological and functional profile. Furthermore, we also suggest that putting must be analyzed rigorously and that the player and the coach are aware that there is indeed a science behind this pendulum motion, and that such 'science' can help to optimize the learning process and workout. Therefore, it is suggested that both coach and athlete act as a 'dynamic system', whereby multilateral work emerges that goes beyond the purely analytical learning and training.

2.5 Summary

This chapter presented a brief overview of one of the most well-known short games ever—*golf putting*. It did so by first describing the pendulum movement and the basis surrounding the learning process associated with it. The first part of the chapter was necessary not only to describe the problem that will be addressed throughout this book, but also to provide the necessary sensitivity to the problems that may arise from this and other similar pendulum-like movements in sport.

We then provided a brief revision of the literature on the evaluation of golf putting. The contributions presented throughout this book are sustained in the state-of-the-art research referred to in Sect. 2.3, especially regarding the proposal of new performance metrics for both the product and process of golf putting.

Nevertheless, before introducing those new mathematical concepts, the next chapter presents the typical experimental setup and all the underlying issues one should consider before starting a research project around golf putting, or any other similar topic.

References

Alexander DL, Kern W (2005) Drive for show and putting for dough? J Sports Econ 6(1):46–60. doi:10.1177/1527002503260797

Beilock SL, Carr TH (2001) On the fragility of skilled performance: what governs choking under pressure? J Exp Psychol 130(4):701–725. doi:10.1110.1037//0096-3445.130.4.701

Beilock SL, Gonso S (2008) Putting in the mind versus putting on the green: expertise, performance time, and the linking of imagery and action. Q J Exp Psychol 61(6):920–932. doi:10.1080/17470210701625626

Brouliette M, Valade G (2008) The effect of putter face grooves on the incipient rolling motion of a golf ball. In D Crews, R. Lutz (eds.) Science and golf V: proceedings of the world scientific congress of golf: 363–368

Couceiro MS, Dias G, Mendes R, Araújo D (2013) Accuracy of pattern detection methods in the performance of golf putting. J Mot Behav 45(1):37–53. doi:10.1080/00222895.2012.740100

Davids K, Button C, Bennett SJ (2008) Dynamics of skill acquisition—a constraints-led approach. Human Kinetics Publishers, Champaign

Dias G, Figueiredo C, Couceiro MS, Luz M, Mendes R (2011) Análise Cinemática do Putt em Jogadores Inexperientes. In: Roseiro L, Neto A (eds) Actas do Congresso. 4.ºCongresso Nacional de Biomecânica. Quinta das lágrimas, Portugal

Dias G, Mendes R, Couceiro MS, Figueiredo C, Luz JMA (2013) "On a ball's trajectory model for putting's evaluation", computational intelligence and decision making—trends and applications, from intelligent systems, control and automation: science and engineering bookseries. Springer, London

Gwyn RG, Patch CE (1993) Comparing two putting styles for putting accuracy. Percept Mot Skills 76:387–390. doi:10.2466/pms.1993.76.2.387

Gwyn RG, Ormond F, Patch CE (1996) Comparing putters with a conventional blade and cylindrically shaped club head. Percept Mot Skills 82:31–34. doi:10.2466/pms.1996.82.1.31

Hume PA, Keogh J, Reid D (2005) The role of biomechanics in maximising distance and accuracy of golf shots. Sports Med 35(5):429–449. doi:10.2165/00007256-200535050-00005

Jonassen DH, Grabowski BL, Hillsdale NJ (1993) Handbook of individual differences, learning and instruction. Lawrence Erlbaum

Karlsen J (2010) Performance in golf putting. Dissertation from the Norwegian School of Sport Sciences

Karlsen J, Smith G (2007) Club shaft weight in putting accuracy and perception of swing parameters in golf putting. Percept Mot Skills 105(1):29–38. doi:10.1016/j.proeng.2010.04.044

Karlsen J, Smith G, Nilsson J (2008) The stroke has only a minor influence on direction consistency in golf putting among elite players. J Sports Sci 26(3):243–250. doi:10.1080/02640410701530902

Mackenzie SJ, Evans DB (2010) Validity and reliability of a new method for measuring putting stroke kinematics using the TOMI1 system. J Sports Sci 28(8):1–9. doi:10.1080/02640411003792711

Magill RA (2011) Motor learning and control: concepts and applications. McGraw-Hill, New York

Mendes PC, Dias G, Mendes R, Martins F, Couceiro MS, Araújo D (2012) The effect of artificial side wind on the serve of competitive tennis players. Int J Perform Anal Sport 12(17):546–562. ISSN 1474-8185 (online)

Newell KM (1986) Constraints on the development of coordination. In: Wade MG, Whiting HTA (eds) Motor development in children: aspects of coordination and control. martinus nijhoff, Boston

Newell KM, Corcos DM (1993) Issues in variability and motor control. In: Newell KM, Corcos DM (eds) variability and motor control. Human Kinetics Publishers, Champaign

Nilsson J, Karlsen J (2006) A new device for evaluating distance and directional performance of golf putters. J Sports Sci 24(2):143–147. doi:10.1080/02640410500131225

Orliaguet JP, Coello Y (1998) Differences between actual and imagined putting movements in golf: a chronometric analysis. Int J Sport Psychol 29:157–169. ISSN 0047-0767

Pelz D (1990) The long putter. The Pelz Report 1:3

Pelz D (2000) Putting bible: the complete guide to mastering the green. Publication Doubleday, New York

Rose D (1997) A multilevel approach to the study of motor learning and control. Allyn & Bacon, Boston

Schöllhorn W, Mayer-Kress G, Newell KM, Michelbrink M (2008) Time scales of adaptive behavior and motor learning in the presence of stochastic perturbations. Hum Mov Sci 28(3):319–333. doi:10.1016/j.humov.2008.10.005

Singer R (2002) Pre-performance state, routines, and automaticity: what does it take to realize expertise in self-paced events? J Appl Sport Psychol 24(4):359–375. ISSN 1543-2904 (online)

Vickers JN (2004) The quiet eye: it's the difference between a good putter and a poor one. Golf Digest 96:32–44. ISSN 0017-176X

Chapter 3
Setting up the Experimental Design

Abstract The past decades have seen technological advances and new analysis tools that allow comprehensive interpretation of human movement patterns. As movement patterns are being analyzed with improved technology, subtle individualities or signature patterns of movement have been identified. These technological advances open a window of opportunity to further study and augment understanding in several fields, including sport science, that benefit from a multidisciplinary approach (biomechanics, engineering, mathematics, motor control) (Dias et al. On a ball's trajectory model for putting's evaluation. Computational intelligence and decision making—trends and applications, from intelligent systems, control and automation: science and engineering bookseries 2013). This chapter may be seen as a guide for setting up the experimental design for research purposes. Several methodological and technological alternatives will be presented and compared. Moreover, a case study provided by the authors and applied to golf putting in the laboratory context will be introduced as an example. Note that the same case study will be used throughout the book. New methods for performance analysis of golf putting have been suggested, focusing on individual kinematic analysis, rather than the traditional pooling of group data, such as in Couceiro et al. (J Motor Behav 45:37–53, 2013) and Dias et al. (On a ball's trajectory model for putting's evaluation. Computational intelligence and decision making—trends and applications, from intelligent systems, control and automation: science and engineering bookseries 2013). The purpose was to understand the relevant changes resulting from the interaction between the athlete's characteristics and the surrounding context by analyzing the motor behavior profiles as measured by the individual kinematic strategies. It is also important to mention that the use of new technologies involves a more profound approach than the traditional linear techniques, which mainly consider product measurements/variables that are a result of the movement (through average, standard deviation or variation coefficient), something that does not allow analysis of the movement itself (Harbourne and Stergiou JNPT Am Phys Ther 89:267–82, 2009).

Keywords Golf putting · Task · Procedures · Instrumentation · Performance

© The Author(s) 2015
G. Dias and M.S. Couceiro, *The Science of Golf Putting*,
SpringerBriefs in Applied Sciences and Technology,
DOI 10.1007/978-3-319-14880-9_3

3.1 Setup

This section explains how one should proceed to prepare and build an experimental setup to analyze golf putting. Note that although we will focus on the golf putting case study, many of the choices presented here may easily be put into practice for other pendulum-like motion and certain ballistic-like actions, such as the tennis serve. This section considers the insights from previous publications, such as Couceiro et al. (2013) and Dias et al. (2013), refining them to a more detailed and tutorial description, and indicating how one can choose the sample (participants), define the task, the procedures and the relevant variables.

3.1.1 Participants

The choice of the participants is of the utmost importance as there is a wide difference between motor execution from novices (who are still learning how to perform a given movement) and from experts (which already completely dominate the movement). When preparing a scientific study, the aim is to ensure the replicability of the obtained results. In spite of this, one must establish the object of analysis and clearly define the criteria considered for sample selection. One of the most important criteria is the level of a player's expertise which will have an influence on dependent and independent variables that will be handled in the research. For instance, Couceiro et al. (2013) analyzed the performance of golf players by comparing their putting patterns through classification methods. By analyzing that work, or any other related work regarding golf putting or any other action, one can easily conclude that novice players do not usually present any sort of playing patterns, i.e., their actions are more chaotic, even random, and present a higher level of variability. Bearing this idea in mind, the authors chose a sample of expert players with a handicap <15, with >10 years of golf practice and participation in national golf tournaments. This is the criteria that any other research should meet to confirm the results obtained by Couceiro et al. (2013).

Besides the level of expertise, other requirements may also influence the final outcome of the research. Namely, in the same study, players were aged 32 ± 10 years, volunteers, male, and right-handed (Couceiro et al. 2013). Although age may not be a critical factor on golf putting performance, researchers will still have to state the average value and standard deviation to improve the level of repeatability of the results. To be even more demanding in the description of the sample, all participants should have legal capacity and competence to participate voluntarily in the investigation. Thus, all athletes are informed in advance that they are participating freely in the study and that they also have complete freedom to withdraw at any time without any penalty. Furthermore, in similar studies, one should state that the study is conducted in accordance with the code of ethics of the proposing institution and the recommendations of the *Declaration of Helsinki on human research.*[1] In other words, performers should not have suffered from any physical or mental disability, being previously informed by written consent that the research does not cause any kind of damage to their physical and mental integrity.

3.1.2 Task and Procedures

Besides describing the sample, the characteristics of the task, the structure of the motor task and the experimental procedures considered should be thoroughly described in the scientific research.

Defining and manipulating the characteristics of the task is important to obtain consistent results with the existing state of the art, while at the same time, it should offer a realistic and challenging context to players. For instance, a very simple task may discourage an expert athlete who already has experience in golf putting, thus resulting in a performance that does not describe his/her real capabilities. On the other hand, a very demanding task may have the opposite effect and discourage an inexperienced athlete who fails to achieve the desired objectives of the researcher. Although obvious, the misconception of the task is very common in scientific research and contaminates the entire study (Guadagnoli and Lee 2004).

The same can be said about the general structure of the motor task. However, in general, this topic has been thoroughly studied by several researchers within the field of motor control (Tani 2005; Davids et al. 2008; Magill 2011), in which particular attention was given to the *contextual interference effect*. This phenomenon is related to the degree of interference that occurs from the learning and performance of a motor skill. Within the several studies on the contextual interference effect in golf putting, such as Tani (2005) and Magill (2011), researchers have identified three possible ways to organize and handle the task (Table 3.1).

The theoretical assumptions that support the contextual interference effect point toward better results in the acquisition phase using a block practice over the alternatives (for in experienced golfers). In contrast, whenever the idea is to retain or transfer the learning, a random practice seems to result in higher levels of motor performance (expert golfers) (Battig 1966; Shea and Morgan 1979; Dias and Mendes 2010).

However, authors such as Porter and Magill (2005) or Dias and Mendes (2010) argue that it may be beneficial to promote a gradual increase of the contextual interference in both learning and evaluation of putting, in order to achieve positive effects, especially in the transfer phase, where it is required to produce a movement similar to the previous one, but with different changes in their temporal characteristics or intensity, such as hitting the ball in a curvilinear trajectory towards the hole.

This hybrid organization means that the practice sessions begin with structured blocks, being followed by series and subsequently by random practice conditions.

---

[1] http://en.wikipedia.org/wiki/Declaration_of_Helsinki.

Table 3.1 Types of motor practice in learning golf putting

Types of motor practice	Distance to the hole: 1, 2 and 3 m
Blocks	1, 1, 1, 1, 2, 2, 2, 2, 3, 3, 3, 3
Series	1,2,3, 1,2,3, 1,2,3, 1,2,3
Random	1,3,2, 3,2,3, 2,1,3, 2,3,2

Nevertheless, any of these studies quantitatively confirm this assumption, thus making it, so far, unequivocally inconclusive and worthy of further exploitation.

Adducing further contributions to these theoretical assumptions, and although it is not considered as a classical hypothesis to explain the contextual interference effect, the 'hypothesis' that was formulated by Guadagnoli and Lee (2004), denoted as challenge point, allows better understanding of the influence of having different levels of complexity in the learning process of motor skills. Under this premise, one can speculate that it is necessary to create an optimal learning/adapting level, such that an inexperienced golfer can adapt to the complexity of golf putting. As previously stated, a low level of complexity may not result in significant contextual interference effects, as it may not motivate the athlete. In contrast, a high level of complexity hardly results in high levels of motor proficiency, as it may discourage the athlete to persist (Guadagnoli and Lee 2004).

Following this idea, we note that authors such as Bjork (1994, 1999), argue that increasingly gradual contextual interference can promote an efficient outcome and stabilize the levels of this effect. Nevertheless, it is concluded that the characteristics of the task and the level of expertise of the athlete are both variables that one needs to consider when designing practice for motor learning in the context of golf putting (Dias and Mendes 2010). For instance, in a previous study by Dias et al. (2014), the main objective was to investigate the adaptation to external constraints and the effects of variability in a golf putting task. The results show that the players changed some parameters to adjust to the task constraints (slope and putting distance), namely the duration of the backswing phase, the speed of the club head and the acceleration at the moment of impact with the ball. Hence, the effects of different golf putting distances in the no-slope condition on different kinematic variables suggest a linear adjustment to distance variation that was not observed during the slope condition. Moreover, the data also indicate that the speed of impact on the ball (process variable) is the one showing a stronger correlation with magnitude of the radial error (product variable), making that variable the best single predictor of golf putting performance.

3.1.3 Variables

Research variables are very important to analyze the sample under a given domain. Moreover, they also determine the type of instruments, organization and experimental procedures that are adopted. As we have seen earlier, it is necessary to know the state of the art in that domain and understand the most typical product variables related to the final result of the operation (e.g., the number of balls that enter the hole), as well as the most typical process variables related to the player's performance (e.g., duration of the movement). During the conception of the experimental setup, one needs to go further and contextualize the *independent* and *dependent variables* under consideration.

Independent variables are intentionally manipulated by the researcher and may assume several states throughout the experiments. In the analysis of golf putting, the independent variables can match the type of structure of the motor task, which may be translated into different types of induced constraints, such as the distance to the hole (e.g., striking the ball at 1, 2, 3 and 4 m to the hole in a straight path) or the addition of a slope between the trajectory of the ball and the hole.

Dependent variables are those that the researcher intends to assess. Typically, the research focuses on understanding how dependent variables are affected by the independent variables. In a study involving golf putting, the dependent variables are related with the performance of the athlete, not only in terms of product, but also in terms of process. They may, for instance, include the distance between the final position of the ball and the hole (radial error), as well as the duration, amplitude, maximum velocity and maximum acceleration of each phase (backswing, downswing, ball impact and follow-through).

3.2   Instrumentation

The instruments used to analyze a given sport movement need to be adapted to the purpose of the research as well as the movement. For instance, while high speed cameras (>200 Hz) may be necessary for a ballistic-like motion, such as the tennis serve, golf putting may be analyzed with a much lower frame rate (e.g., 30 Hz). Nevertheless, and depending on the level of detail one may need (e.g., analyze the tilt of the putter head), high resolution cameras may still be required.

Cameras or any other instruments (e.g., accelerometers) may provide a detailed feedback about the motion and the final result of the action that may not be ensured, at least accurately, by the naked eye.

3.2.1   Traditional Cameras

In golf putting, as with most pendulum-like movements, most of the analysis can be achieved with the use of bidimensional tracking systems, such as a single traditional camera, e.g., complementary metal oxide—semiconductor cameras. This simple method allows one to assess valuable and accurate information about the putting action or ball trajectory which, on its own, can provide important feedback to athletes and coaches (Neal and Wilson 1985; Couceiro et al. 2012). New ever-improving and low-cost cameras have allowed successive important break-throughs in most sports, including golf putting, by providing more information (high rates and high resolution) about action patterns and how these evolve over time (Neal et al. 2007; Dias et al. 2013).

Dias et al. (2013) used a photography camera (Casio Exilim/High Speed EX-FH25) featuring filming capabilities of up to 210 Hz at a resolution of 480 × 360 pixels, with a lens of 26 mm focal length. One camera was used to capture the movement, while another one, exactly the same, was used to capture the ball's trajectory (see camera 2). Figure 3.1 illustrates the experimental setup adopted by the authors. This kind of illustration is helpful so other researchers may replicate the study. Furthermore, it provides an overall idea of the setup.

The main disadvantage regarding traditional cameras is the use of computa-tionally complex, and sometimes unreliable, detection and estimation methods. For instance, Couceiro et al. (2012) presented a strategy of detection, estimation and classification applied to the golf putting context. The detection method consisted of a simple geometry and color match. Afterwards, the estimation method consisted of fitting the trajectory of the object (golf putter) with a mathematical model, in which its parameters were optimized comparing several techniques in terms of compu-tational complexity and memory. Five different estimation techniques were studied, applied and compared, namely gradient descent, pattern search, downhill simplex, particle swarm optimization (PSO) and Darwinian particle swarm optimization (DPSO). Results confirmed the superior performance of the DPSO method (Couceiro et al. 2012).

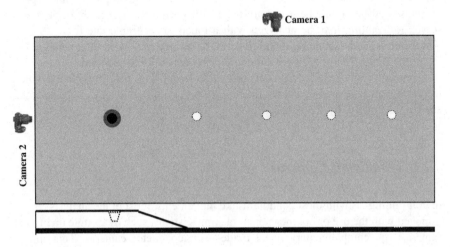

Fig. 3.1 *Top* and *upper* view of the experimental setup from Dias et al. (2013)

Afterwards, several classification algorithms were considered to extract the unique signatures of each player by considering the parameters previously optimized. The work tested several classifiers, from the most traditional methods, such as the linear discriminant analysis (LDA) or the quadratic discriminant analysis (QDA), to state-of-the-art approaches, such as naive Bayes (NB) and least-squares support vector machines (LS-SVM). The classification methods were compared through analysis of the confusion matrix and the area under the receiver operating characteristic curve (AUC), and the SVM presented an overall better performance.

As one may assume, the simplicity, and low-cost of using a single monocular camera is shortly transformed into a series of other complications that most sport scientists try to avoid using, for instance, semi-autonomous or completely manual tracking software, e.g., *AnaMov* and *Tacto*. Even the most state-of-the-art computer vision methods tend to fail in unstructured or outdoor scenarios, due to occlusions, light variations, and other phenomena.

Although many ever-improving strategies have been proposed over the past few years to tackle such issues, monocular cameras are still more reliable in the laboratory workspace where one can benefit from several landmarks previously employed in the environment. To overcome these limitations, computer vision methods have been crossing the bridge from two-dimensional (2D) to three-dimensional (3D) by benefiting from depth cameras.

3.2.2 3D Depth Cameras

The development of depth cameras for pose estimation brought new opportunities for human motion analysis (Shotton et al. 2013). Among the multiple depth camera choices, stereo vision and time-of-flight technologies are perhaps the most widely used. The success of depth cameras has been successively attested by commercial systems that estimate full body poses for computer games, hand poses for action interfaces, or capture detailed head motions for facial animation (Ye et al. 2013).

The revolution of depth cameras into research started with the appearance of the *Microsoft Kinect.*[2] Before it came onto the market, depth imaging was costly (>1,000 Euros). However, *Kinect* was released for the game industry at a cost of approximately 250 Euros. Such low-cost property soon led engineers and roboticists to adapt its features to other domains, namely for simultaneous localization and mapping in robotics (Endres et al. 2012) or, more specifically in the context of this book, skeleton pose tracking in biomechanical engineering and sport sciences (Fernández-Baena et al. 2012).

---

[2] http://www.microsoft.com/en-us/kinectforwindows/.

Despite these achievements, computer vision techniques applied to 3D imaging require a higher computational effort than their 2D counterparts. For instance, most state-of-the-art depth computer vision techniques are able to track >48 DOF of the human skeleton up to 30 Hz. Considering that a camera of approximately the same cost can acquire approximately 200 Hz, this presents a drawback that one should consider during the design phase of the experimental setup. For instance, although the *Kinect* sensor, as any other similar depth camera such as the *ASUS Xtion,*[3] could be easily deployed to acquire a player's kinematics while performing golf putting, for faster actions, such as the tennis serve, it would be unfeasible. Moreover, similar to their 2D counterparts, depth cameras are also susceptible to lighting conditions, thus constraining their adequate use to controlled and preferably indoor environments.

3.2.3 Motion Capture Suits

Motion capture suits have been more and more in vogue for human movement analysis (Garafalo 2010). Notwithstanding the range of different suits, these sections only focus on inertial sensor-based systems. Any alternatives to this technology (e.g., stereophotogrammetry-based systems) fall within the limitations described in the previous section, since they depend on the use of cameras which restrict their use within the laboratory workspace.

Inertial sensor-based motion capture suits, such as the *MTx* units from *Xsens* solutions,[4] benefit from multisensor fusion techniques applied to inertial measurement units (IMU), comprising of gyroscopes, accelerometers and magnetometers, so as to accurately estimate the 3D position of each joint. These suits are generally equipped with wireless transmitter units that send synchronized data to a data logger communicating with a computer. Therefore, such instrumental analysis can be adopted by athletes in order to allow them to mostly concentrate on the task, without constraints imposed by camera-based solutions (e.g., light conditions). Moreover, contrary to the previous alternatives, motion capture suits are completely portable, thus offering the possibility to be applied outdoors and in the field context.

However, all these benefits do not come without a cost. The state-of-the-art *Xsens* suit, being currently used not only for research but also in film and game industries, costs approximately 60,000 Euros. Although one can find cheaper alternatives on the market, such as the *3DSuit* solution[5] costing approximately 25,000 Euros, the cost is still significantly higher than the previous two alternatives.

---

[3] http://www.asus.com/Multimedia/Xtion_PRO/.

[4] http://www.xsens.com/.

[5] http://3dsuit.com/.

3.2.4 Sport-Specific Devices

All the previous devices can be used for data retrieval in most sport modalities. Nevertheless, some sport-specific devices are also available on the market, such as *Sony's Smart Tennis Sensor*[6] and *Ingeniarius' InPutter.*[7] The main reason behind the development of these instrumented devices is to maintain the ecological validity of the overall setup, without additionally constraining the athlete with unrealistic situations (e.g., laboratory setup, full-body suit, etc.).

In the golf putting context, the recently developed *InPutter* seems to be the one device that completely maintains the ecological validity of the setup. Although other products have been developed to study and improve the putting performance, such as the *SAM PuttLab,*[8] the *InPutter* is the only one that does not require any auxiliary hardware or that is confined to laboratory experiments. In brief, *InPutter* is an engineered golf putter designed for research, analysis and training purposes. By benefiting from an internal IMU sensor and wireless technology, it is able to retrieve the most relevant golf putting process variables, namely the putter's trajectory over time, speed, duration and amplitude of each phase, as well as the impact force on the ball. The system additionally includes a heart rate monitor interface compatible with *Polar* transmitters.[9] As *InPutter* does not require any camera systems, markers, or system infrastructure, and given its robustness, weight and design which is similar to other traditional golf putters, it can be used in both indoor and outdoor environments. Additionally, *InPutter* is an internet-connected product that automatically connects to the *Ingeniarius Cloud,*[10] thus allowing real-time debugging and monitoring over the internet.

The *InPutter* was developed by *Ingeniarius, Lda.*, a private company, in which the authors of this book played a vital role in its development; the first author of this book, Dr. Gonçalo Dias, was the key consultant and researcher during the development of the product, having the main responsibility to maintain its ecological validity. The co-author of this book, Dr. Micael Couceiro, was the co-founder of *Ingeniarius, Lda.* and the supervisor of the engineering development of *InPutter*. Given the positive feedback from many golf players, namely from the European Pitch and Putt Champion Hugo Espirito Santo, *InPutter* is currently being used not only within the research context, but in real-life golf training. Table 3.2 summarizes several golf-specific devices and their features.

---

[6] http://www.theverge.com/2014/1/20/5326558/sony-smart-tennis-sensor-price-and-availability.

[7] http://www.ingeniarius.pt/inputter.

[8] http://www.samputtlab.com/.

[9] http://www.polar.com/en.

[10] http://cloud.ingeniarius.pt/.

Table 3.2 Benchmark of golf putting devices available on the market

		IPING[a]	3BaysGSA[b]	TOMI PRO[c]	SAMPuttLab[d]	InPutter[e]
EXPORT					✓	✓
3D VISUALIZATION						✓
HEARTRATE						✓
IMPACT VELOCITY			✓	✓	✓	✓
PEAK ACCELERATION					✓	✓
OVERALL AMPLITUDE			✓	✓	✓	✓
OVERALL DURATION		✓	✓	✓	✓	✓
FACE ANGLE		✓	✓	✓	✓	✓
DECLINATION ANGLE						✓
IMPACT	DURATION					✓
IMPACT	FORCE					✓
IMPACT	POSITION			✓	✓	✓
FOLLOW THROUGH	DURATION				✓	✓
FOLLOW THROUGH	AMPLITUDE					✓
DOWN SWING	DURATION	✓	✓		✓	✓
DOWN SWING	AMPLITUDE				✓	✓
BACK SWING	DURATION	✓	✓		✓	✓
BACK SWING	AMPLITUDE				✓	✓

[a] http://www.ping.com/fitting/iping.aspx
[b] http://www.3bayslife.com/gsa/home.php
[c] http://tomi.com/
[d] http://www.samputtlab.com/
[e] http://ingeniarius.pt/inputter.html

3.3 Practical Implications

The evolution in technology, especially over the past decade, resulted in some major advances in golf putting analysis. The feedback from this analysis, being either qualitative or quantitative, is of high importance as it provides a deeper knowledge around putting. Although the presented information revolves around golf putting, it can be optimized and used to analyze other sports. Actually, a higher degree of performance analysis may allow a completely new understanding about the individual singular properties of each player. Using this pertinent information, one can adjust the training programs to the player's specificities.

3.4 Summary

In summary, experimental procedures are of the utmost importance in the analysis of sport movements such as golf putting. In this sense, before starting operationally to study a given action such as putting, it is very important to answer the following questions:

(1) Where will the task be performed (laboratory context, indoors, outdoors, real competition, etc.)?
(2) What will the research focus on (product and process variables, physiological variables, etc.)?
(3) What information is given and how it will be provided to the participants (verbally, by video, by demonstration, or mixed information)?
(4) What materials will be used and manipulated by the participants (putter and golf ball), clearly stating whether they will always use the same materials throughout the study or materials will vary according to the purpose of the task?
(5) What is the full range of the trials under study (distance to the hole, pose of the player, etc.)?

Answering these questions will help define the necessary instrumentation hardware and software that one will need for a particular case study. Moreover, it is advisable to conduct a preliminary study to fully evaluate the defined setup and determine if the answers to the previous five questions are adequate.

For instance, the work by Dias et al. (2014) evaluated the setup for expert golfers a posteriori, in works such as Couceiro et al. (2012) and Dias et al. (2013), by considering three novice players (namely the research team). That preliminary study allowed gaps for some of the answers provided to the five questions above to be filled. For example, although the number of conditions was adequately defined, each condition comprised 30 trials, thus resulting in 270 trials for the whole study. All novice athletes, although novice and considerably different in terms of motor performance than experts, showed extreme levels of fatigue and their performance significantly dropped after approximately 200 puttings. Therefore, each condition was resized to 20 trials each, thus resulting in 180 trials for the whole study.

That preliminary study also allowed a rethink about adequate data acquisition hardware. To further improve the accurate identification of the ball impact phase, the putter was equipped with an accelerometer. Nevertheless, the use of two

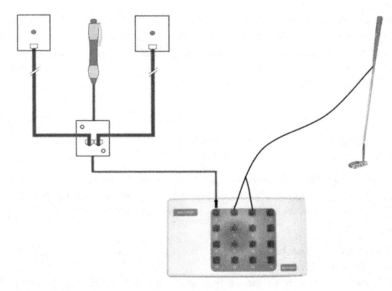

Fig. 3.2 Experimental setup developed for data synchronization (adapted from Dias et al. 2010)

cameras (Fig. 3.1) and one accelerometer led to the development of a trigger mechanism so as to synchronize the instrumentation setup (Fig. 3.2).

Nevertheless, any good and even mathematically solid methodology may fail if the setup is not thoroughly planned. The performance evaluation may be hindered due to an inadequate experimental design. It is, therefore, noteworthy that although the next two chapters present multiple strategies to evaluate the performance of golf putting, the applicability of such methods depends highly on having an adequate experimental design.

References

Battig WF (1966) Facilitation and interference. In: Bilodeau EA (ed) Acquisition of skill. Academic Press, New York

Bjork RA (1994) Memory and metamory considerations in the training of human beings. In: Metcalfe J, Shimamura A (eds) Metacognition: knowledge about knowing. MIT Press, Cambridge

Bjork RA (1999) Assessing our own competence: heuristics and illusions. In: D Gopher, Koriat A (eds) Attention and performance XVII. Cognitive regulation of performance: interaction of theory and application. MA: MIT Press, Cambridge

Dias G, Mendes R, Luz M, Couceiro, MS, Figueiredo C (2010) Procedimentos e análise de dados aplicados ao estudo do putting. Boletim da SPEF 35(4):47–58. ISSN 1646-8775

Couceiro MS, Dias G, Martins FML, Luz JM (2012) A fractional calculus approach for the evaluation of the golf lip-out. Signal Image Video 6(3):437–443. doi:10.1007/s11760-012-0317-1

Couceiro MS, Dias G, Mendes R, Araújo D (2013) Accuracy of pattern detection methods in the performance of golf putting. J Motor Behav 45(1):37–53. doi:10.1080/00222895.2012.740100

Davids K, Button C, Bennett SJ (2008) Dynamics of skill acquisition—a constraints-led approach. Human Kinetics Publishers, Champaign

Dias G, Mendes R (2010) Efeitos do contínuo de níveis de interferência contextual na aprendizagem do "putt" do golfe. Rev bras cienc esporte 24(4):545–553. doi:10.1590/S1807-55092010000400011

Dias G, Mendes R, Couceiro MS, Figueiredo C, Luz JMA (2013) "On a Ball's trajectory model for putting's evaluation", computational intelligence and decision making—trends and applications, from intelligent systems, control and automation: science and engineering bookseries. Springer Verlag, London

Dias G, Couceiro MS, Barreiros J, Clemente FM, Mendes R, Martins FM (2014) Distance and slope constraints: adaptation and variability in golf putting. Mot Control 18(3):221–243. doi:10.1123/mc.2013-0055

Endres F, Hess J, Engelhard N, Sturm J, Cremers D, Burgard W (2012) An evaluation of the RGB-D SLAM system. In: 2012 IEEE International Conference on robotics and automation (ICRA) pp 1691–1696

Fernández-Baena A, Susín A, Lligadas X (2012) Biomechanical validation of upper-body and lower-body joint movements of kinect motion capture data for rehabilitation treatments. In: 4th International Conference on intelligent networking and collaborative systems (INCoS) pp 656–661

Garafalo P (2010) Development of motion analysis protocols based on inertial sensors, Ph.D. thesis on Bioengineering, University of Bologna

Guadagnoli MA, Lee TD (2004) Challenge point: a framework for conceptualizing the effects of various practice conditions in motor learning. J Motor Behav 36(2):212–224. doi:10.3200/JMBR.36.2.212-224

Harbourne RT, Stergiou N (2009) Movement variability and the use of nonlinear tools: principles to guide physical therapist practice. JNPT Am Phys Ther 89(3):267–282

Magill RA (2011) Motor learning and control: concepts and applications. McGraw-Hill, New York

Neal RJ, Wilson BD (1985) 3D kinematics and kinetics of the golf putting. Int J Biomech 1(3):221–232. ISSN (online): 1543-2688

Neal RJ, Lumsden R, Holland M, Mason B (2007) Body segment sequencing and timing in golf. In: Jenkings S (ed) Annual review of golf coaching 2007. Multi-Science Publication, Brentwood

Porter JM, Magill RA (2005) Practicing along the contextual interference continuum increases performance of a golf putting task. J Sport Exerc Psy 27:S-124. ISSN. 0895-2779

Shea JB, Morgan RL (1979) Contextual interference effects on the acquisition, retention and transfer of motor skill. J Exp Psychol Learn Mem Cogn 5(2):178–187. doi:10.1037/0278-7393.5.2.179

Shotton J, Sharp T, Kipman A, Fitzgibbon A, Finocchio M, Blake A, Moore R (2013) Real-time human pose recognition in parts from single depth images. Commun ACM 56(1):116–124

Tani G (2005) Comportamento motor: aprendizagem e desenvolvimento. São Paulo, Guanabara

Ye M, Zhang Q, Wang L, Zhu J, Yang R, Gall J (2013) A survey on human motion analysis from depth data. In: time-of-flight and depth imaging. sensors, algorithms, and applications Springer, Berlin Heidelberg pp. 149–187

Chapter 4
Performance Metrics for the Putting Product

Abstract By taking advantage of the available technology, researchers have focused their attention on the development of new methodologies to study human movement (Dias et al. Motor Control 18:221241, 2014). Several studies have analyzed the variables influencing the performance of golf putting, mainly focusing on the measures of product performance (Dias et al. computational intelligence and decision making—trends and applications, from intelligent systems, control and automation: science and engineering book series, 2013; Couceiro et al. J Motor Behav 45:3753, 2013). The majority of this research has been carried out in a laboratory context, where the green is usually emulated using a carpet (e.g., Delay et al. Hum Mov Sci 16:597619, 1997; Coello et al. Int J Sport Phychol 31:2446, 2000), with circular targets, or holes of different sizes (Dias and Mendes Rev Bras Cienc Esporte 24:545553, 2010; Dias et al. computational intelligence and decision making—trends and applications, from intelligent systems, control and automation: science and engineering book series, 2013). Most of the methodologies adopted by these studies describe the quantification of the motor performance error based on the final location of the ball in relation to the center of the hole, commonly known as radial error (Couceiro et al. J Motor Behav 45:3753, 2013; Dias et al. computational intelligence and decision making—trends and applications, from intelligent systems, control and automation: science and engineering book series, 2013). Despite their usefulness, this and similar techniques are not sufficient to describe a player's performance as a whole, bearing in mind that they only consider the 'cause and effect' linear actions resulting from an athlete's actions or match situations. For instance, one needs to better understand the meaning of having a golf ball finishing before, after, or in the vicinity of the hole, and to what extent this is truly meaningful and important for the putting performance. Moreover, it should be noted that current research is scarce around this topic and does not clarify in any way, the difference between a golf ball that stays in the ±90° lines towards the hole, or another that stays in the ±180° lines towards the hole. In other words, the literature does not provide any theoretical support for this research question, thus reinforcing the evaluation methodology proposed here. This chapter introduces alternative methods to further assess and understand the performance of golf putting in terms of product, i.e., in terms of end result. It is

© The Author(s) 2015
G. Dias and M.S. Couceiro, *The Science of Golf Putting*,
SpringerBriefs in Applied Sciences and Technology,
DOI 10.1007/978-3-319-14880-9_4

noteworthy that we do not aim at replacing the more traditional measures, but on complimenting the information they may provide.

Keywords Golf putting · Product variables · Performance · Fuzzy approach

4.1 Binary Evaluation

One way to evaluate golf putting accuracy consists of counting the number of balls that enter the hole. In this book, we denote this metric as the *binary error*. Despite not being referred to in the literature as such, this metric may be consubstantiated in theoretical assumptions that cover the discipline of engineering, notably the concepts inherent in binary logic (Couceiro et al. 2012). By applying this logic to the golf putting performance, it is possible to evaluate a player's performance by quantifying the number of times that the ball entered the hole by comparison with the total number of trials of motor practice. For example, according to binary logic (e.g., values 0 and 1), if the player placed the ball in the hole, 1 would be regarded as *success* and if the ball did not enter the hole the value attributed would be 0, i.e., *failure*. In other words, the values obtained do not have intermediate values (Couceiro et al. 2012).

In the present study, this metric is represented by:

$$\eta_B = \frac{1}{N} \sum_{i=1}^{N} n_i \qquad (4.1)$$

where N is the total number of trials and n_i is the binary value that represents putting *success*, i.e., if the ball enters the hole $n_i = 1$, otherwise, $n_i = 0$.

4.1.1 Evaluation

To evaluate this and the subsequent metrics, the sample and procedures described in the previous section were adopted, in which 10 expert golfers were studied. In that sense, Fig. 4.1 shows the players' performance using the binary error metric. As one may observe, player 1 shows the best performance; he obtained a success rate of 83 %, i.e., he hit 83 % of the balls. Players 4 and 9 follow, with player 2 having the worst performance.

This metric is directly related with the accuracy of a given player; the more balls he/she are able to place in the hole, the higher his/her game score. However, considering the metric η_B, if the player scores as many times as he fails, his evaluation will be 0.5, not taking into account the value of the error obtained when

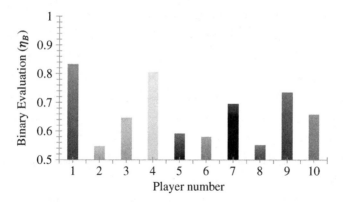

Fig. 4.1 Binary evaluation of players

he failed. The result is a value that does not take into account the consistency of a player's performance, as he may obtain a high evaluation despite having trials with high error values.

4.2 Radial Error

In recent studies focusing on the analysis of golf putting, the *radial error* has been used to examine the product measures derived from the motor performance of players (Dias et al. 2014). Through the analysis of longitudinal, lateral and radial errors, the same authors obtain quantitative values resulting from the final position of the ball as to the center of the hole. When the player hits the hole, his error is considered (zero) 0 in the components of longitudinal, lateral and radial errors. In this context, the metric of performance of a player will be defined by the arithmetic mean of the radial error obtained in each trial:

$$\mu_R = \frac{1}{N}\sum_{i=1}^{N}\varepsilon_i, \tag{4.2}$$

where ε_i is the radial error of trial i. Through analysis of metric μ_R, we conclude that the higher the value, the worse the putting accuracy performance.

The radial error is obtained through the following expression:

$$\varepsilon_i = \sqrt{(\varepsilon_i^x)^2 + (\varepsilon_i^y)^2}, \tag{4.3}$$

which results from the application of the *Pythagorean theorem*, where the legs of the triangle are defined by the lateral error ε_i^x and the longitudinal error ε_i^y, and the hypotenuse is defined by the radial error ε_i (Fig. 4.2).

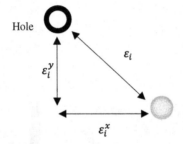

Fig. 4.2 Representation of the three measured errors (adapted from Couceiro et al. 2012)

However, on representing an absolute value, the metric μ_R makes it difficult to compare multiple athletes. One of the ways to overcome this, and as presented in a previous study (Couceiro et al. 2012), consists of applying a measure of ration based on a maximum radial error value of ε_{max}, which depends on the evaluative and normative criteria (e.g., handicap, green limits and the player's distance to the hole), always being higher than or equal to the radial error ε_i for any trial i of any player, i.e., $\varepsilon_i \leq \varepsilon_{max}$ $\forall i$. In this way, the relative metric η_R is obtained through the expression:

$$\eta_R = \frac{1}{N} \sum_{i=1}^{N} \left(1 - \frac{\varepsilon_i}{\varepsilon_{max}} \right). \tag{4.4}$$

4.2.1 Evaluation

As before, let us observe the radial error performance for each one of the 10 expert golfers. In this particular study, a maximum threshold value was found for player 8, $\beta = \varepsilon_{max} = 3{,}778$ mm (Fig. 4.3).

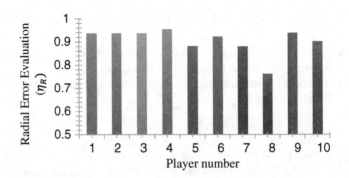

Fig. 4.3 Radial error evaluation of the players in the three experimental studies

The lowest radial error is found for player 4, who comes nearest to 1. Players 9 and 1 closely follow player 4. Finally, player 8 clearly shows the worst performance. This metric allows an 'analogical' evaluation of the putting accuracy (i.e., it is not solely represented by the *success* or *failure* of this movement). However, it should be noted that a trial that promotes the placement of the ball inside the hole tends to be considered slightly better than one which results in a ball close to it. Furthermore, this evaluation does not take into account all the 'dynamics' of the putting performance, since the radial error may lead to an erroneous evaluation, thus concealing the results obtained. As previously mentioned, unlike other movements (e.g., javelin throw), in golf putting the lateral error ε_i^x may not carry the same 'weight' as the longitudinal error ε_i^y in the calculation of the metric of the putting performance evaluation.

4.3 Argument Error Evaluation

The concept adopted in this metric dates back to the pioneering studies conducted by Isaac Newton, which paved the way to research into the polar coordinates as we know them today. From an operational point of view, a semi-straight line which is based at the origin, i.e., point (0, 0), and a point in the cartesian plane (x, y) may be represented in the polar plane as a modulus (i.e., length of the semi-straight line as to the origin) and an argument (i.e., the angle that this semi-straight line forms with a line parallel to the abscissa x) (Stewart 2007). In order to evaluate putting accuracy performance, we propose a new evaluation metric, denoted *argument error*, based on the methodology previously described, by considering radial error as the modulus (i.e., absolute value) of an error (which was denoted as the polar error) represented in the polar coordinate system as follows:

$$\dot{\varepsilon}_i = \varepsilon_i \angle \theta_i, \qquad (4.5)$$

where the argument θ_i is obtained through the conversion of the cartesian coordinate system $\left(\varepsilon_i^x, \varepsilon_i^y\right)$ into the polar coordinate system $(\varepsilon_i, \theta_i)$, where θ_i is obtained by using the arctangent variation *atan2* function that takes the quadrant into account.

Through conversion of cartesian coordinates into polar coordinates we obtain the quadrant where the ball was placed at the end of each trial of motor practice and the circular positions around the origin (i.e., hole), as shown in Fig. 4.4. It should be noted that the y-axis is aligned with the line defined by the exit point of the ball (i.e., where the impact moment of the putter with the ball takes place) and the center of the hole. On the other hand, the x-axis is perpendicular to the y-axis and aligned with the center of the hole.

On considering the polar coordinate system, we observe that the distance which balls 1 and 2 are from the hole is the same, i.e., $\varepsilon_1 = \varepsilon_2$. However, ball 2 is situated in the first quadrant, close to the 0° line with $\theta_2 = 10°$, while ball 1 is situated in the second quadrant, close to the 90° line with $\theta_1 = 100°$.

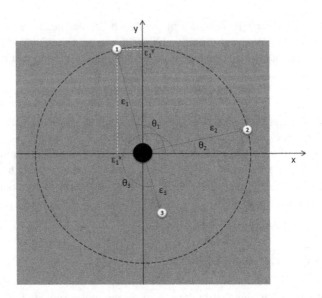

Fig. 4.4 Graphic representation of the cartesian and polar coordinate system based on the *center* of the hole (adapted from Couceiro et al. 2012)

The question here is to know which of the two situations represents a better performance at the putting performance level. According to the metrics defined by the expressions (4.2) and (4.4), both would represent the same accuracy, i.e., neither of them entered the hole and both are at the same distance from it (Fig. 4.4).

To better answer the previously formulated question, putting analysis is used in the laboratory and field context. Although not explicitly referred to in some studies (Pelz 2000; Mackenzie and Evans 2010), a player's performance tends to be considered 'worse' when the balls are close to the line that separates the first from the fourth quadrant and to the line that separates the second from the third quadrant. As mentioned above, the lateral error ε_i^x proves to be more 'critical' than the positive longitudinal error ε_i^y. This means that through the analysis of radial error ε_i, in combination with the argument of radial error θ_i, it is possible to determine whether the accuracy of a given trial was better than another.

According to previously described studies (Pelz 2000; Mackenzie and Evans 2010), it is preferable to obtain angles close to $\pm 90°$ than to angles close to $0°$ or $180°$, by making the putting accuracy represented by ball 1 higher than the putting accuracy of ball 2. On the other hand, it may also be observed that the putting accuracy represented by ball 3 will be higher than the putting accuracy resulting in the two other positions represented by balls 1 and 2, since $\varepsilon_3 < \varepsilon_1$ and being found in the fourth quadrant with $\theta_3 = 280°$.

In order to compare the accuracy performance through the argument error, as performed for the radial error, we developed a novel relative metric defined as follows:

$$\eta_\theta = \frac{1}{N} \sum_{i=1}^{N} \left(1 - \frac{||\theta_i| - 90|}{90} \right). \tag{4.6}$$

4.3.1 Evaluation

Figure 4.1 shows a player's performance by only considering the argument error. One may observe that player 1 shows the best performance, being nearer to 1. Players 7 and 8 follow, with player 2 having the worst performance (Fig. 4.5).

To further improve the feasibility of the proposed error, a Spearman's rank correlation was conducted between the handicap of 10 golf players and their η_θ obtained in a preliminary study. This study consisted of 30 trials performed by each player at 4 m from the hole without any constraints. A Spearman's rank correlation coefficient of $\rho_s = -0.4893$ was obtained, showing a decreasing monotonic trend between a player's handicap and his accuracy performance through the angular position of the balls. In other words, one can consider that as the handicap decreases, the putters finish closer to the $\pm 90°$ lines.

Nevertheless, the fact that there are multiple evaluation metrics to determine a player's performance makes their selection process complex. Due to the 'dynamics' of the putting performance, it may not be sufficient to consider each evaluation metric independently. It is thus extremely important to find a way to evaluate a player's performance and simultaneously ponder the binary metric (it entered or did not enter the hole) and the radial error (if it did not enter the hole, its proximity to it), and the argument error (in case it entered the hole, the angle location of the ball).

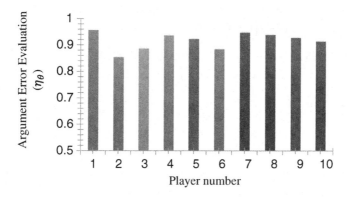

Fig. 4.5 Argument error evaluation of the players in the three experimental studies

Consequently, it is based on the fuzzy approach, introduced in the next section, that we will evaluate the performance with regard to accuracy that contemplates various factors associated with a game of golf.

4.4 Fuzzy Approach

It is possible through fuzzy logic to transform quantitative variables into qualitative ones by describing not only the 'total' error obtained by the player but also the extent of his failure in the same trial. In order to do so, this research considers three entries for the diffuse system, which are defined by the three types of errors previously introduced (i.e., binary error, radial error and argument error).

The fuzzy system (i.e., membership functions and rules) was formulated based on conscious and subconscious human knowledge acquired from 2,700 putting trials demonstrated by 10 expert golf players (Couceiro et al. 2012). Based on the accuracy and precision of the product variable (i.e., final position of the ball) it was possible to assess the relationship between the inputs and output of the fuzzy system (Fig. 4.6).

The membership function of the binary error will be represented by a unitary crisp membership function, as shown in (4.7) and in Fig. 4.4. That is, value 0 will be ascribed if the player fails, i.e., shot without *success* (*failure*) and value 1, in case he puts the ball in the hole, i.e., shot with *success* (Fig. 4.7).

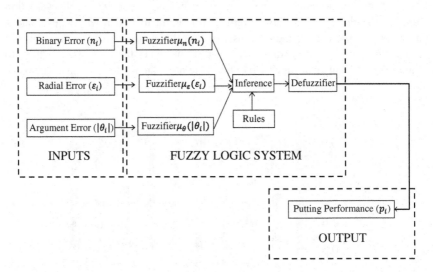

Fig. 4.6 The fuzzy logic system for evaluating putting performance (adapted from Couceiro et al. 2014)

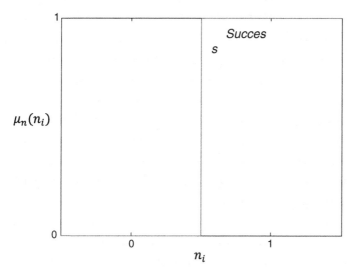

Fig. 4.7 Binary error membership function (adapted from Couceiro et al. 2014)

$$\mu_n(n_i) = \begin{cases} 0, & n_i = 0 \\ 1, & n_i = 1 \end{cases}. \tag{4.7}$$

It should be highlighted that this method has been used in the analysis of dynamic systems, in particular at the level of studies involving artificial intelligence. For more details, see the work by Bart (1992) and Couceiro et al. (2012).

The membership function of the radial error may be represented as a special case of a triangular membership function (Fig. 4.8).

The smaller the radial error, the closer the ball will be to the hole. Based on Eq. (4.4), this function may be represented as follows:

$$\mu_\varepsilon(\varepsilon_i) = \begin{cases} 1 - \frac{\varepsilon_i}{\beta}, & \varepsilon_i \leq \beta \\ 0, & \varepsilon_i > \beta \end{cases}, \tag{4.8}$$

where parameter β is equal to $\varepsilon_{\max}$ (cf. Eq. 4.4) which, as previously mentioned, may be related with the green's size or the highest radial error recorded in all the trials under study. The membership function of the argument error (i.e., angle of the ball to the hole) will be defined by a generalized bell-shaped membership function, shown in Fig. 4.6. This function has one more parameter than the typical Gaussian function used in membership functions:

$$\mu_\theta(|\theta_i|) = \frac{1}{\left[1 + \frac{(|\theta_i| - c)}{a} \right]^{2b}}, \tag{4.9}$$

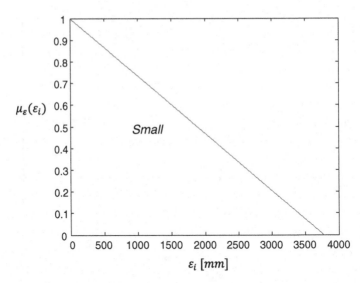

Fig. 4.8 Radial error membership function with $\varepsilon_{max} = 3,778$ mm (adapted from Couceiro et al. 2014)

where the parameters a, b and c correspond to the width, slope and center of the curve, respectively. Consequently, we decided to use the absolute value of the argument error, $|\theta_i|$, as input, considering the negative angles (i.e., third and fourth quadrants) as having the same evaluation as the positive angles (i.e., first and second quadrants). Parameters a, b and c are initialized, considering that a ball situated on the $|90|°$ line (i.e., *right* angle) has the maximum performance value as to the argument error, whereas a ball near to the $0°$ or $180°$ line (i.e., between the first and fourth quadrant and between the second and third quadrant, respectively) represents a shot with a lower performance when compared with the argument error. Therefore, considering Eq. (4.6) and using *MatLab*'s *Fuzzy Logic Toolbox*, we obtain $a = 25$, $b = 1.5$ and $c = 90$ (Fig. 4.9).

For defuzzification, we used Mamdani's implication with the lowest (i.e., first) of the maximums. Basically, two single diffuse *IF–THEN* rules to allow the classification of a given putting trial were considered:

IF n_i is *succes* THEN p_i is *accurate with weight 1*
ELSE-IF ε_i is *small* AND $|\theta_i|$ is *right* THEN p_i is *accurate with weight 0.8*.

The first rule is the most relevant to classify the putting success, as it represents the performance metric of the golf player, contemplating whether or not the ball entered the hole.

If it enters, i.e., $n_i = 1$, it will not be necessary to analyze the radial error (which will be zero), and the shot will be classified as *accurate*, i.e., $p_i = 1$.

On the other hand, if the ball does not enter the hole, i.e., $n_i = 0$, the radial error will have to be pondered. Nevertheless, it was decided that the weight of this

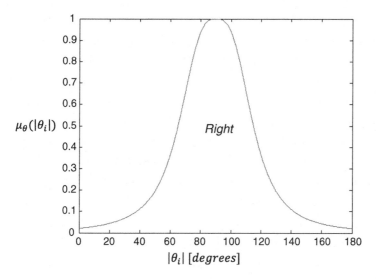

Fig. 4.9 Argument error modulus membership function (adapted from Couceiro et al. 2014)

second rule would be 0.8, since as of the moment that the player does not hit the hole the putting will be considered to have, at best, an accuracy of 80 %, i.e., $p_i = 0.8$.

As to the radial error, we know that the higher it is, the worse the player's performance will be with regard to product variable. However, this relationship will only be linear if the argument error remains constant. If the argument error modulus comes close to the limits of 0° and 180°, the putting will be considered to have a lower performance and, consequently, to have lower accuracy. The question here is to know:

How accurate will the putting be when the radial error argument varies?

This relationship cannot be considered linear because of the features inherent in the execution of the putting. For example, considering that a given putting trial was unsuccessful, i.e., $n_i = 0$, and had a *small* radial error of 10 cm. If the argument error modulus is close to 90°, i.e., $\mu_\theta(|\theta_i|) \approx 1$, the shot is considered to have a certain accuracy. However, assuming that the argument error modulus is close to 0° or 180°, having the same radial error of 10 cm, it is possible to evaluate the putting with the same level of accuracy.

On the other hand, we will assume that instead of obtaining a radial error of 10 cm we have an error of 100 cm. In this situation, the argument error modulus close to 90° will result in a putting performance which is far higher than the performance of the same shot with an argument error modulus close to 0° or 180°.

This means that the *AND* connective cannot be considered as usual, i.e., minimum or product between the radial error membership function and the argument error modulus membership function. Therefore, a new *AND* connective is then

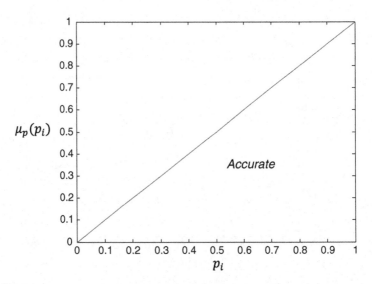

Fig. 4.10 Consequent function of the putting performance (adapted from Couceiro et al. 2014)

proposed to relate both membership functions in order to maintain the relationship previously referred to:

$$AND = \mu_\varepsilon(\varepsilon_i) \times \mu_\theta(|\theta_i|)^{1-\mu_\varepsilon(\varepsilon_i)}. \tag{4.10}$$

As a result, the lower the radial error, i.e., $\mu_\varepsilon(\varepsilon_i) \approx 1$, the lower the influence of the argument error modulus, and vice versa.

Finally, the consequent function is defined as follows:

$$\mu_p(p_i) = \begin{cases} p_i, & \varepsilon_i \geq 0 \\ 0, & \varepsilon_i < 0 \end{cases}, \tag{4.11}$$

The value $\mu_p(p_i)$ is the performance evaluation with regard to the accuracy of trial i (Fig. 4.10).

Figure 4.10 presents the output surface of the fuzzy system without considering the input of the binary error (since the other input variables only matter when $n_i = 0$). It is noteworthy that a greater influence of the argument error is observed for larger values of the radial error, i.e., as the radial error increases the curvature of the surface also increases.

In order to understand the evaluation metric proposed, we shall consider the following example. Two golf players made two trials each, having been recorded that player 1 hit the first trial, i.e., $\varepsilon_1^1 = 0$, but in the second trial the radial error obtained was the highest radial error of all the trials made, with an argument error of $0°$, i.e., $\dot{\varepsilon}_2^1 = \varepsilon_{\max} \angle 0°$. Player 2 failed both trials with a radial error 10 times lower

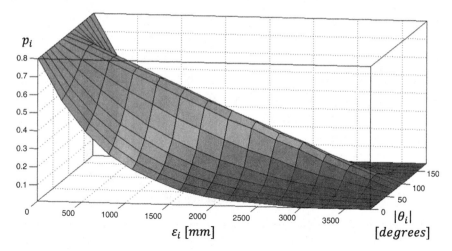

Fig. 4.11 Mapping from radial and argument error to putting performance with $\varepsilon_{\max} = 3{,}778\,\text{mm}$ (adapted from Couceiro et al. 2014)

than $\varepsilon_{\max}$, having obtained an argument error of $90°$ in the first trial and an argument error of $0°$ in the second trial, i.e., $\dot{\varepsilon}_1^2 = \frac{\varepsilon_{\max}}{10} \angle 90°$ and $\dot{\varepsilon}_2^2 = \frac{\varepsilon_{\max}}{10} \angle 0°$.

In this situation the scenario would be the following—as regards the binary error, player 1 would have a performance higher than that of player 2, as he hit a ball when compared to player 2, who hit none, i.e., $\eta_B^1 = 0.5 > \eta_B^2 = 0$. As regards the radial error, player 2 would be the best with $\eta_R^2 = 0.9 > \eta_R^1 = 0.5$. Both players would present the same performance as to the argument error with $\eta_\theta^1 = \eta_\theta^2 = 0.5$.

Given the context, and using fuzzy evaluation, we will have $\mu_p^1 = 0.5 > \mu_p^2 = 0.36$. This means that player 1 had a 14 % higher performance than player 2 (Fig. 4.11).

In the following section, the performance of various expert players is compared using the fuzzy-based approach presented here.

4.4.1 Evaluation

To evaluate players using fuzzy logic, we benefited from the fuzzy inference system (*FIS*) editor (*FIS Editor*) of the *Fuzzy Logic Toolbox*. Complementing the information previously presented, Fig. 4.12 and Table 4.1 show the players' performance relying on diffuse information.

In study 1, player 1 achieves the best performance, followed by players 4 and 9. Player 8 has the worst performance. In study 2, player 3 achieves the best performance, followed by players 10 and 4. The data also show that faced with a

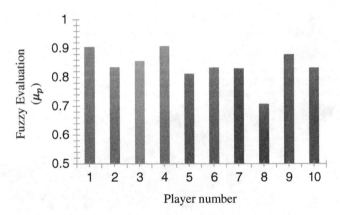

Fig. 4.12 Fuzzy evaluation of players in the three experimental studies (adapted from Couceiro et al. 2014)

constraint (ramp/slope), the accuracy performance of player 1 tends to decrease considerably when compared with the performance of player 4. Player 8 shows the worst performance. In study 3, player 5 shows the best performance, placing approximately 75 % of the balls in the hole. Players 6 and 7 follow player 5. Again, player 8 shows the worst performance. Player 4 obtained the best performance in the three studies (Fig. 4.2), followed by players 1 and 9. Finally, player 8 shows the worst performance (Total row from Table 4.1). Table 4.1 depicts the overall result obtained using the fuzzified approach throughout the three experimental studies.

Considering the results from Table 4.1, it is important to retrieve as much information as possible about the putting execution so as to understand how this method can be applied to the sports training context. If we only consider the aim of the golf game, which involves placing the ball in the hole with as few shots as possible, player 1 would be considered as the best performer (cf. binary error) followed by players 4 and 9. However, if one considers the performance evaluation with regard to fuzzy logic accuracy, i.e., the simultaneous analysis of the binary error, the radial error and the argument error, then player 4 is the one presenting the best performance, followed by players 1 and 9.

One can conclude that, although being suitable to evaluate situations that involve subjectivity, vagueness and imprecise information, the metric relies on the experience of selecting the adequate membership function. Therefore, expert knowledge about the task is required in order to validate the proposed rules and membership functions. In other words, fuzzy systems need expert experience to strengthen the decision rules and to handle imprecise value in its reasoning.

Table 4.1 Fuzzy performance obtained by players throughout the three experimental studies (adapted from Couceiro et al. 2014)

Study	S01	S02	S03	S04	S05	S06	S07	S08	S09	S10
E1_1 m	1.0000	1.0000	1.0000	1.0000	1.0000	1.0000	1.0000	1.0000	0.9787	1.0000
E1_2 m	0.9847	0.8947	0.9517	0.9757	0.8590	0.8290	0.8807	0.7917	0.9640	0.8923
E1_3 m	1.0000	0.7850	0.7700	0.9860	0.7260	0.7703	0.8510	0.7657	0.9530	0.8633
E1_4 m	1.0000	0.7407	0.8050	0.9367	0.8047	0.7340	0.8503	0.4550	10.000	0.6267
E1	0.9962	0.8551	0.8817	0.9746	0.8474	0.8333	0.8955	0.7531	0.9739	0.8456
E2_2 m	0.9413	0.9320	0.9610	0.9647	0.8400	0.9507	0.9277	0.8820	0.9493	0.9477
E2_3 m	0.9273	0.8820	0.9847	0.9777	0.8323	0.8757	0.8270	0.6777	0.8587	0.9713
E2_4 m	0.8657	0.8380	0.9033	0.8953	0.7133	0.8320	0.6153	0.5763	0.8307	0.9327
E2	0.9114	0.8840	0.9497	0.9459	0.7952	0.8861	0.7900	0.7120	0.8796	0.9506
E3_ang1	0.8237	0.7523	0.7843	0.6923	0.7763	0.7407	0.7710	0.6747	0.7447	0.6823
E3_ang2	0.6157	0.7027	0.5533	0.7553	0.7590	0.7827	0.7490	0.5550	0.6437	0.5753
E3	0.7197	0.7275	0.6688	0.7238	0.7677	0.7617	0.7600	0.6148	0.6942	0.6288
Total	0.9065	0.8364	0.8570	0.9093	0.8123	0.8350	0.8302	0.7087	0.8803	0.8324

Legend E1 (study 1); E2 (study 2); E3 (study 3); 1 m (1 meter); 2 m (2 meters); 3 m (3 meters); 4 (4 meters); ang1 (angle 1, left); ang2 (angle 2, right); S01 (player 1); S02 (player 2); S03 (player 3); S04 (player 4); S05 (player 5); S06 (player 6); S07 (player 7); S08 (player 8); S09 (player 9); S10 (player 10)

4.5 Practical Implications

The new recent technologies developed to understand human movement have inevitably boosted golf putting research. By merging such technologies with the computation of new metrics, one can better understand the reason behind the end result of this motor skill which emerges as an important practical implication for players, coaches and researchers. This approach combines biomechanics, mathematics, sports and engineering, fostering a multidisciplinary and integrative point-of-view on how we should investigate the ball's trajectory over the hole. Moreover, the several methods presented here provide a set of robust responses inherent in the motor execution of the golfers, from novices to experts. It is our belief that these metrics can be applied to other golf movements, such as chip and pitch, in order to investigate the true 'mechanics' behind them.

4.6 Summary

The several metrics presented in this chapter bring implications to the area of sports training since they aim to provide a deeper understanding of a player's flaws (Kolev et al. 2005). These approaches are important mainly in a coaching perspective to avoid overusing standard metrics that lack relevant information about a given action (Dias and Mendes 2010). For instance, although player 1 was the best performing player most of the time, his overall performance significantly dropped when considering other product variables. In that sense, this multidisciplinary approach brings us a step nearer to understanding golf putting and the necessary information required during training and competition. Meanwhile, fuzzy logic has practical applications in other individual and team sports (e.g., tennis, football, basketball) that can benefit from this type of fuzzified metric with both quantitative and qualitative information, being mainly useful to follow the performance trend of athletes' motor behavior. Such techniques are equally effective to assess how the athlete can stabilize his/her performance by exploring different levels of variability and complexity. Moreover, this approach is extremely useful to measure the performance fluctuations and irregularities of both novices and experts, as well as to assess their individual motor skill characteristics and profiles (Couceiro et al. 2012).

Operationally, this study introduces new evaluation metrics that are relevant in sports so as to measure the performance of athletes in laboratory and real-life situations, for both teaching and learning. Specifically in the context of golf putting, these metrics show that it is possible to devise a 'memory' that objectively provides a trend to a player's performance during the execution of the task. In this case, the player is able to monitor his/her motor progress and correct errors resulting from the putting performance. Moreover, it also allows quantification of the result of the action and the direction of the error in the context of training and competition (Couceiro et al. 2012).

Despite these achievements, the metrics presented in this chapter also present some limitations. As they revolve around product variables, they do not consider the motor skill of the athlete. In spite of this, the next chapter introduces traditional and non-traditional perspectives around process variables.

References

Bart K (1992) Neural networks and fuzzy systems: a dynamical systems approach to machine intelligence. Pren-tice-Hall, Englewood Cliffs

Coello Y, Delay D, Nougier V, Orliaguet JP (2000) Temporal control of impact movement: the "time from departure" control hypothesis in golf putting. Int J Sport Psychol 31(1):24–46. ISSN 0047-0767

Couceiro MS, Dias G, Martins FML, Luz JM (2012) A fractional calculus approach for the evaluation of the golf lip-out. Signal Image Video P 6(3):437–443. doi:10.1007/s11760-012-0317-1

Couceiro MS, Dias G, Mendes R, Araújo D (2013) Accuracy of pattern detection methods in the performance of golf putting. J Motor Behav 45(1):37–53. doi:10.1080/00222895.2012.740100

Couceiro MS, Martins FML, Clemente F, Dias G, Mendes R (2014) On a fuzzy approach for the evaluation of golf players. Maejo Int J Sci Technol 8(01):86–99. ISSN 1905-7873

Delay D, Nougier V, Orliaguet JP, Coello Y (1997) Movement control in golf putting. Hum Mov Sci 16 (5):597–619. doi:10.1016/S0167-9457(97)00008-0

Dias G, Mendes R (2010) Efeitos do contínuo de níveis de interferência contextual na aprendizagem do "putt" do golfe. Rev Bras Cienc Esporte 24(4):545–553. doi:10.1590/S1807-55092010000400011

Dias G, Mendes R, Couceiro MS, Figueiredo C, Luz JMA (2013) On a Ball's trajectory model for putting's evaluation, computational intelligence and decision making—trends and applications, from intelligent systems, control and automation: science and engineering book series. Springer, London

Dias G, Couceiro MS, Barreiros J, Clemente FM, Mendes R, Martins FM (2014) Distance and slope constraints: adaptation and variability in golf putting. Motor Control 18(3):221–243. doi:10.1123/mc.2013-0055

Kolev B, Chountas P, Petrounias I, Kodogiannis V (2005) An application of intuitionistic fuzzy relational databases in football match result predictions. Adv Soft Comput 2:281–289

Mackenzie SJ, Evans, DB (2010) Validity and reliability of a new method for measuring putting stroke kinematics using the TOMI1 system. J Sports Sci 28(8):1–9

Pelz D (2000) Putting bible: the complete guide to mastering the green. Publication Doubleday, New York

Stewart J (2007) Calculus: early transcendental. McGraw Hill, New York

Chapter 5
Performance Metrics for the Putting Process

Abstract Although most of the traditional research around sport science is centered on the product variables, many researchers have been working toward a better understanding of the process measurements of motor execution. By studying those variables, one may further understand the reasons behind the stability and variability of the final outcome (i.e., the product variables previously presented). In spite of this, several authors, such as Delay et al. (1997), Coello et al. (2000), Hume et al. (2005), Couceiro et al. (2013) and Dias et al. (2013) have proposed methodologies to study process variables in golf putting during each of its phases (cf. Chap. 2), giving particular attention to the position, velocity and acceleration of the putter. Most of the current research on this subject focuses on specific properties of the process variables, such as amplitude, period, maximum or minimum values, etc. This chapter will start by presenting the insights regarding variables provided in the literature. Despite the useful information provided by those process variables, the difficulty remains in proposing an adequate analysis methodology encompassing the overall motor execution of the athlete. This is still considered an open challenge since, as opposed to the product variables previously presented, most of the process variables, either related to golf putting or not, are time-variant, i.e., they are classified as a time series. Those variables are directly related with human movement and, as biological processes, the analysis should consider the overall information over time. However, this sort of tool, mostly of a non-linear nature, requires specialized knowledge on engineering and mathematics. Nevertheless, after introducing the most traditional research around putting process variables, this chapter will delineate a methodology to bring the science of golf to a new level of understanding.

Keywords Golf putting · Process variables · Performance · Time series analysis · Non-linear methods · Measures and statistics

© The Author(s) 2015
G. Dias and M.S. Couceiro, *The Science of Golf Putting*,
SpringerBriefs in Applied Sciences and Technology,
DOI 10.1007/978-3-319-14880-9_5

5.1 Measures and Statistics

All the research within the context of golf putting classifies this action as a complex motor ability that significantly changes from player to player, depending on his/her individual characteristics and profile.

In one of the first most complete scientific studies aiming to understand the golf putting process variables, Coello et al. (2000) retrieved a set of measures that would be the starting point to not only understand motor performance but also to compare different players. In that study, the authors were able to verify that athletes presented a ball impact velocity of ~ 1.25, 1.56 and 1.89 ms^{-1} for a distance to the hole of 2, 3 and 4 m, respectively. From this preliminary study, it was possible to observe a direct and still non-linear relationship between the ball impact velocity and the distance to the hole. They were also able to verify that athletes depicted a downswing amplitude of 200, 237 and 280 mm for a distance to the hole of 2, 3 and 4 m, respectively, following a similar relationship as observed for impact velocity (Table 5.1).

Table 5.1 Average (M), standard deviation (SD) and variation coefficient (VC %) values for golf putting process variables with and without a slope constraint (adapted from Dias et al. 2014a)

Putting process variables	Values[a]	Practice condition #1 without slope constraint			Practice condition #2 with slope constraint		
		2 m	3 m	4 m	2 m	3 m	4 m
Backswing/downswing[a] amplitude (DS) (mm)	M	171	178	186	261	234	231
	SD	36	26	29	37	35	43
	VC %	21	15	16	14	15	19
Follow-through amplitude (FT) (mm)	M	292	365	414	413	493	500
	SD	42	44	52	47	52	79
	VC %	14	12	13	11	11	16
Impact velocity on the ball (VI) (ms^{-1})	M	1.14	1.28	1.41	1.6	1.81	1.72
	SD	0.18	0.15	0.24	0.17	0.22	0.28
	VC %	16	12	17	11	12	16
Backswing duration time (BS) (ms)	M	459	461	559	463	442	578
	SD	97	98	140	76	111	166
	VC %	21	21	25	16	25	29
Downswing duration time (DS) (ms)	M	290	294	298	300	277	325
	SD	59	48	64	43	49	85
	VC %	20	16	22	14	18	26
Follow-through duration time (FT) (ms)	M	437	469	493	446	477	499
	SD	91	79	100	89	104	106
	VC %	21	17	20	20	22	21
Maximum acceleration of the putting (AM) (ms^{-2})	M	5.48	5.45	6.04	7.36	8.46	7.15
	SD	0.62	0.33	0.56	0.82	1.47	0.45
	VC %	11	6	9	11	17	6

Delay et al. (1997) compared the putting performance of expert athletes with inexperienced players, concluding that the latter would present a larger ball impact velocity of 1.5 ms^{-1} versus 1.3 ms^{-1}. In that same study, the authors were able to conclude that expert athletes, as opposed to inexperienced players, were able to maintain velocity peaks of the action uniformly throughout the trials while varying the distances to the hole.

The duration of the putting action, namely for each of its phases, was also one of the measures targeted by researchers. Coello et al. (2000) found that the downswing duration would be 269, 271 and 279 ms for distances to the hole of 2, 3 and 4 m, respectively, confirming the gradual increment of the downswing duration with the distance to the hole. Delay et al. (1997) confirmed this result to some extent by concluding that the downswing duration would last between 261 and 269 ms for distances between 2 and 4 m, respectively, and that the duration would slightly increase for inexperienced players. This study also showed that expert athletes tend to start the putting execution closer to the ball and end further away from it compared to inexperienced players (i.e., with a larger follow-through). In line with these results, Karlsen (2003) concluded that the downswing duration would be stable regardless of the expert athlete, as no statistical differences were found between different athletes. Moreover, Karlsen et al. (2008) later verified that too large a downswing (superior to 370 ms) could lead to a negative effect on the execution of the action, especially when players took too long to prepare for that phase.

Although most of the literature has addressed the measures and properties of the process behind the execution of putting, the data is scattered or incomplete. Table 5.1 summarizes the experiments carried out by Couceiro et al. (2013) and Dias et al. (2014a), emphasizing the most relevant state-of-the-art measures. The sample consisted of ten adult male golfers aged 33.80 ± 11.89 years, who were volunteers, right-handed and expert players (10.82 ± 5.4 handicap). Each player performed 30 putts at distances of 2, 3 and 4 m (90 trials in practice condition #1). The participants also performed 90 trials, at the same distances with a constraint imposed by a slope (practice condition #2).

The general overview presented in Table 5.1 presents the average, standard deviation and variation coefficient values for the dependent variables associated with the process of golf putting. The following subsections give a more thorough explanation about these variables focusing on their relevance for each phase of golf putting.

5.1.1 Amplitude

The results of this study indicated that the total golf putting amplitude tends to increase with golf putting distances in both practice conditions. Overall, the downswing amplitude between distances was significantly different ($F_{(2.897)} = 5.102$; $p = 0.006$; $\eta^2 = 0.011$; power = 0.822), as well as the follow-through amplitude

$(F_{(2.897)} = 91.696; p = 0.001; \eta^2 = 0.170; \text{power} = 1.00)$. However, there is a clear increment of the backswing/downswing amplitude in longer golf putting distances (e.g., 3 and 4 m) in practice condition #1 that was not observed in practice condition #2. The average backswing/downswing amplitude without the slope is 178.33 mm, against the average amplitude of 242.00 mm with the slope. The introduction of a slope is responsible for an average increment of 63.67 mm in the amplitude of the movement. It is important to note that the slope has a moderate impact on the backswing/downswing amplitude and a strong effect on the follow-through amplitude (111.67 mm).

In that sense, the inter-trial variability of the downswing amplitude is higher in the slope condition, but only at longer distances (3 and 4 m)—the mean variability increased 30 % at the 3 m distance and 46 % at the 4 m distance. The variability of the downswing amplitude at the distance of 2 m in practice condition #1 (±35.40 mm) and at the distance of 4 m (±41.41 mm) in practice condition #2 is higher than in the remaining conditions. The inter-trial variability of the downswing amplitude is higher in the slope condition, but only at longer distances (3 and 4 m)—the mean variability increased 30 % at the 3 m distance and 46 % at the 4 m distance.

5.1.2 Duration

The effect of distance was mainly observed in the duration of the backswing and of the follow-through components. On the contrary, the duration of the downswing phase remains similar among distances and conditions. In practice condition #1, the maximum difference between distances was 8 ms, while in practice condition #2 the difference between the 3 m distance and the 4 m distance was 48 ms. Overall, the duration of the golf putting followed a similar trend than the one observed for the amplitude, except for practice condition #2 when the players were 2 m from the hole. In practice condition #1, a one-way ANOVA revealed significant differences between distances—backswing duration $(F_{(2.897)} = 47.512; p = 0.00; \eta^2 = 0.096;$ power = 1.00), follow-through duration $(F_{(2.897)} = 12.010; p = 0.00; \eta^2 = 0.026;$ power = 0.995), but not for the downswing duration $(F_{(2.897)} = 0.819; p = 0.441;$ $\eta^2 = 0.002; \text{power} = 0.191)$. In practice condition #2, all distances had a significant effect on the duration of the phases—backswing $(F_{(2.897)} = 59.284; p = 0.001;$ $\eta^2 = 0.117; \text{power} = 1.00)$, downswing $(F_{(2.897)} = 22.684; p = 0.001; \eta^2 = 0.048;$ power = 1.00) and follow-through $(F_{(2.897)} = 8.912; p = 0.001; \eta^2 = 0.019;$ power = 0.973).

The variability of the backswing duration is higher at the longest golf putting distance, in both practice conditions (±131.6 and ±154.32 ms). In the slope condition (practice condition #2), the backswing duration consistently increases with distance —there was a 42 % increment of inter-trial variability from 2 to 3 m and a 52 % increment from 3 to 4 m. In general, the variability of the follow-through duration increases with distance but not necessarily with the slope. Apparently, the follow-through variability is less dependent of the two contextual variables under analysis.

5.1.3 Velocity

The club head speed, or velocity, at the moment of contact with the ball was systematically adjusted by players according to the distance to the hole for practice condition #1, varying from 1.14 to 1.41 ms^{-1}, with an approximate speed increment of 0.14 ms^{-1} for each meter of distance to the hole, representing a 12 % increase of the club head speed for 1 m of extra distance. In practice condition #2, the average club head speed at the three distances was 1.71 ms^{-1}, against 1.28 ms^{-1} in the no slope condition—the slope was responsible for an overall 34 % increase of the club head speed. Statistically speaking, in practice condition #1, the impact velocity was significantly different based on the distance to the hole ($F_{(2.897)} = 69.973$; $p = 0.00$; $\eta^2 = 0.135$; power = 1.00), as well as in practice condition #2 ($F_{(2.897)} = 23.362$; $p = 0.001$; $\eta^2 = 0.05$; power = 1.00). Remarkably, the 4-m putt with a slope did not register the highest club head speed, as expected. The higher values of impact velocity under practice condition #2 were found when players were 3 m from the hole. In summary, the 4-m putt with a slope registered the longest duration of backswing, downswing and follow-through (578, 325 and 499 ms, respectively) but not the highest impact velocity or acceleration.

In that sense, the variability of the speed of impact of the ball tends to increase with distance but only at the longer distances (3 and 4 m). At these distances the introduction of a slope also tends to increase inter-trial variability (50 % at 3 m and 18 % at 4 m).

5.1.4 Evaluation

Figure 4.9 represents the linear tendency (i.e., straight-line equation) of the ratio between the duration of the *downswing* and duration of the *backswing* for each practice condition. The optimized parameter was the mean square error (MSE) (status: 0.22 for practice condition #1, and 0.18 for practice condition #2) (Fig. 5.1).

As observed, the trend of the line is decreasing, i.e., with a negative slope, inversely dependent with the distance to the hole, regardless of the practice condition. Furthermore, the slope of the linear trend for the second practice condition is slightly inferior to the one for the first practice condition; there are differences between both increases regarding distance to the hole.

The data also show that, regardless of the practice condition, there is an increase of ~ 5 % with an increase of 1 m away from the hole, i.e., a decreasing rate of 5 % per meter. In other words, the duration of the *downswing* tends to decrease 5 % more than the duration of the *backswing* with the distance to the hole. Note that the *downswing* phase takes approximately half the time required for the execution of the *backswing* for a distance of 4 m.

Finally, the results show that the players change some parameters to adjust to the task constraints, namely the duration of the backswing phase, the speed of the club

Fig. 5.1 Linear trend
(straight line equation) of the
ratio between the duration of
backswing and *downswing*,
for each practice condition.
D = distance (2, 3 and 4 m)

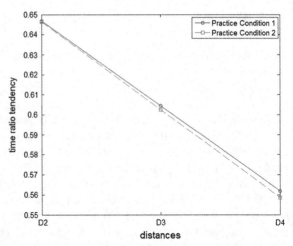

Legend: D = Distance (2, 3 and 4 meters).

head and the acceleration at the moment of impact with the ball. In that sense, the
effects of different golf putting distances in the no-slope condition on different
kinematic variables suggest a linear adjustment to distance variation that was not
observed during the slope condition.

The results of this study indicate that the speed of impact on the ball shows a
stronger correlation with magnitude of the radial error, making that variable the best
single predictor of the golf putting performance. Using the variation coefficient of
the speed impact on the ball as an indicator of the player's stability, we have
established a ranking for the player's performance under the two practice conditions
(Table 5.2).

Table 5.2 Motor stability ranking of players under the two practice conditions

	Rank		Player	Mean radial error	Variable coefficient (VC) (%)
Stability Boundary VC = 24.61 %	Stable	1	8	1,115.15	14.51
		2	10	548.31	17.59
		3	1	312.79	17.73
		4	3	293.85	24.18
		5	6	461.29	24.6
	Unstable	6	7	708.53	24.63
		7	2	322.96	24.89
		8	5	658.96	25.03
		9	9	299.61	28.6
		10	4	245.26	32.28

The stability boundary simply represents the threshold value that separates the stability of the motor execution from the instability in the context of the performance of the golf players under the two practice conditions.

In that sense, the stability boundary is settled as the median of the variation coefficients of the ball impact speed (VC %) with a value of 24.61, thus dividing the group of players into two groups identified as stable and unstable. In this ranking, player golfer 8 is the most stable (VC % = 14.51) while player 4 is the most unstable player (VC % = 32.28).

5.2 Time Series Non-linear Analysis

To describe the variability of human motor behavior in the context of sport performance, it has been established that nonlinear techniques, such as the approximate entropy and Lyapunov exponent, allow the unraveling of the structure of a mathematical representation of a given sport movement, like golf putting. In spite of nonlinear techniques quantifying the motor performance of athletes through average, standard deviation and coefficient of variation, they take the individual characteristics of players into account and are mostly based upon statistical effects to characterize the learning and training of motor skills (Stergiou et al. 2004). Faced with such arguments, it seems that the problem of 'individuality' in sport is not confined to ideal, linear or standardized techniques. In fact, it has to do with the implementation of a wide variety of exercises that contribute to the self-organization of the motor system.

Non-linear techniques provide qualitative information on the tendency of the motor system by observing different patterns of response. Unlike cognitive theories that support traditional motor control models, which consider variability to be a negative factor for learning, the non-linear perspective shows that 'noise' is necessary to establish new coordinative patterns (Davids et al. 2008). In that sense, 'noise' is considered as random fluctuations that incorporate a certain spectrum of action. Thus, several types of noise are well known in the literature (e.g., pink, white, brown and black). For example, pink noise is related to the study of the human heart rate while white noise can be measured on electromyography signals (Stergiou et al. 2004).

Furthermore, it seems difficult to study the variability that characterizes the motor performance of golfers in putting performance based only upon traditional statistical results (e.g., mean, standard deviation and coefficient of variation), as is common procedure in most studies that have analyzed this movement in the laboratory context, as well as in training and competition (Schöllhorn et al. 2008). Note that this is typically represented by a sequence of data points, typically consisting of successive measurements of the athlete's performance made over a time interval (i.e., time series). From this perspective, non-linear applications can be used in the study of the variability of systems of human movement by complementing classical linear techniques, which are normally used to quantify the performance of motor skills. However, it should be highlighted that this is not about under-rating the important role of linear techniques in the research of systems of human movement, but rather about deepening their study in harmony with non-linear tools (Stergiou et al. 2004; Harbourne and Stergiou 2009).

5.2.1 Preparing Data

In order to evaluate putting under a given specific condition (e.g., of a given player, under a specific practice condition, etc.) one needs to generate a single time series that may represent the overall putting data over time in all the trials under that same condition. In other words, after obtaining the data of each putting one wishes to evaluate as a hole, it is necessary to concatenate the necessary data. By doing so, it would then be possible to numerically calculate the metrics of the non-linear analysis temporally by concatenating the putting trajectory of trials (T) carried out under that same condition.

For instance, Fig. 5.2 depicts 30 trials of a single athlete under a specific practice condition concatenated into a single time series. Generally speaking, the mathematical model from Fig. 5.2 would represent the time series characteristic of a player's movement during a given practice condition. In the case represented in Fig. 5.2, it is possible to confirm that the player presents some variability at the level of putting execution (the amplitude and duration of the movement slightly diverges throughout the trials).

The representation between the golf club position and speed at each instant was used to better understand the movement through its attractor (Fig. 5.3).

Figure 5.3 confirms that the movement is located somewhere between the periodic (circular image around a point) and the chaotic (distortion in amplitude and shape of the image). However, it is difficult to quantify the variability of the player. Using non-linear methods that allow for the characterization of the variability of the

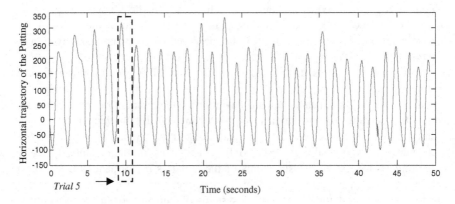

Fig. 5.2 Example of the concatenation of 30 trials (adapted from Dias et al. 2014b)

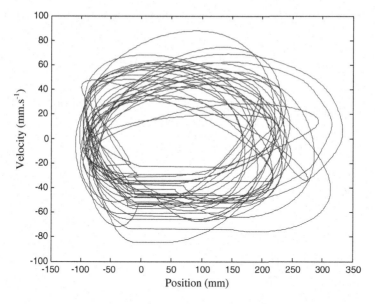

Fig. 5.3 Attractor resulting from the concatenation of 30 trials (adapted from Dias et al. 2014b)

player during a determined practice condition becomes important (Harbourne and Stergiou 2009).

Both approximate entropy and largest Lyapunov exponent will be considered to further understand the variability of golf players. Nevertheless, throughout the years, several different methods have been proposed to calculate both the approximate entropy and the largest Lyapunov exponent (Dias et al. 2014b). The next sections present the chosen approaches based on a preliminary assessment of the related work applied to human movement.

5.2.2 Approximate Entropy

Pincus et al. (1991) described the techniques for estimating the Kolmogorov entropy of a process represented by a time series and the related statistics approximate entropy. In this sense, consider that the whole data of the T trials is represented by a time series as $u(1), u(2), \ldots, u(N) \epsilon \mathbb{R}$, from measurements equally spaced in time, which form a sequence of vectors $x(1), x(2), \ldots, x(N - m + 1) \in \mathbb{R}^{1 \times m}$, defined by:

$$x(i) = [\, u(i) \quad u(i+1) \quad \cdots \quad u(i+m-1)\,] \in \mathbb{R}^{1 \times m}.$$

The parameters N, m and r must be fixed for each calculation. N is the length of the time series (number of data points of the whole series), m is the length of sequences to be compared and r is the tolerance for accepting matches. One can define:

$$C_i^m(r) = \frac{\text{number of } j \text{ such that} \leq r}{N - m + 1},$$ (5.1)

for $1 \leq i \leq N - m + 1$. Defining $d(x(i), x(j))$ for vectors $x(i)$ and $x(j)$, and based on the work of Takens (1981) and (Dias et al. 2014b), it results in:

$$d(x(i), x(j)) = \max_{k=1,2,\ldots,m} [|u(i+k-1) - u(j+k-1)|].$$ (5.2)

From the $C_i^m(r)$, it is possible to define:

$$C_i^m(r) = (N - m + 1)^{-1} \sum_{i=1}^{N-m+1} C_i^m(r),$$ (5.3)

and

$$\beta_m = \lim_{n \to 0N \to \infty} \frac{\ln C_i^m(r)}{\ln r}.$$ (5.4)

The assertion is that for a sufficiently large m, β_m is the correlation dimension. Such a limiting slope has been shown to exist for the commonly studied chaotic attractors. This procedure has frequently been applied to experimental data. Researchers seek a 'scaling range' of r values for which $\frac{\ln C_i^m(r)}{\ln r}$ is nearly constant for large m, and they infer that this ratio is the correlation dimension (Grassberger and Procaccia 1983). Some researchers have concluded that this procedure establishes deterministic chaos (Pincus et al. 1991; Pincus and Singer 1998; Stergiou et al. 2004; Dias et al. 2014b).

The following relationship is defined:

$$\phi^m(r) = (N - m + 1)^{-1} \sum_{i=1}^{N-m+1} \ln C_i^m(r).$$ (5.5)

One can define the approximate entropy as:

$$ApEn(m, r, N) = \Phi^m(r) - \Phi^{m+1}(r).$$ (5.6)

On the basis of calculations that included the theoretical analysis performed by Pincus et al. (1991), a preliminary estimate showed that choices of r ranging from 0.1 to 0.2 of the standard deviation of the data would produce reasonable statistical validity of $ApEn(m, r, N)$. As a consequence, values of approximate entropy close to zero characterize a periodical signal/system of high regularity, low variability and little complexity. Following this line of thought, values of approximate entropy ≥ 1.5, qualify as a signal/system of high variability, low complexity and little regularity (Pincus et al. 1991; Pincus and Singer 1998; Harbourne and Stergiou 2009; Dias et al. 2014b).

5.2.3 Lyapunov Exponent

Using the Lyapunov exponent, it is possible to quantify the sensitivity of initial conditions of dynamical systems. Within the golf context, the spectrum of the Lyapunov exponent can classify the divergence of putting trajectories. This concept relates to the spectrum of the Lyapunov exponent by considering a small n dimensional sphere of initial conditions, in which n is the number of equations used to describe the system (Rosenstein et al. 1993). The Lyapunov exponent may be arranged so that (Dias et al. 2014b):

$$\lambda_1 \geq \lambda_2 \geq \cdots \geq \lambda_n, \tag{5.7}$$

where λ_1 to λ_n corresponds to the most rapidly expanding and contracting principal axes, respectively. Hence, one needs to recognize that the length of the first principal axis is proportional to $e^{\lambda_1 t}$, so that the area determined by the first two principal axes is proportional to $e^{(\lambda_1+\lambda_2)t}$ and the volume determined by the first k principal axes is proportional to $e^{(\lambda_1+\lambda_2+\cdots+\lambda_k)t}$. Therefore, the Lyapunov spectrum can be defined so that the exponential growth of a k-volume element is given by the sum of the k largest Lyapunov exponents. The largest Lyapunov exponent can then be defined by using the following equation, where $d(t)$ is the average divergence at time t and C is a constant that normalizes the initial separation:

$$d(t) = Ce^{\lambda_1 t}. \tag{5.8}$$

In order to improve the convergence (with respect to i), Sato et al. (1987) proposed the following equation:

$$\lambda_1(i, k) = \frac{1}{k \cdot \Delta t} \cdot \frac{1}{M - k} \sum_{j=1}^{M-k} \frac{\ln\left(d_j(i+k)\right)}{d_j(i)}, \tag{5.9}$$

where M is the number of axes being analyzed. The golf putt can be described by analyzing only the horizontal axis, x-axis, $M = 1$. From the definition of λ_1 given in Eq. (5.10), we assume that the jth pair of nearest neighbors diverges approximately at a rate given by the largest Lyapunov exponent (Dias et al. 2014b):

$$d_j(i) \approx C_j e^{\lambda_1(i \cdot \Delta t)}, \tag{5.10}$$

where C_j is the initial separation.

By taking the logarithm of both sides of Eq. (5.12) the following is obtained:

$$\ln d_j(i) \approx \ln C_j + \lambda_1(i \cdot \Delta t). \tag{5.11}$$

Equation (5.13) represents a set of approximately parallel lines (for $j = 1, 2, \ldots, M$), each with a slope roughly proportional to λ_1. The largest Lyapunov exponent is easily and accurately calculated by using a least-squares fitting to the 'average' line defined by:

$$y(i) = \frac{1}{\Delta t \langle \ln d_j(i) \rangle}, \tag{5.12}$$

where $\langle \ln d_j(i) \rangle$ denotes the average of $\ln d_j(i)$ over all values of j. This process of averaging is the key to calculating accurate values of λ_1 using small, noisy data sets (Dias et al. 2014b).

The calculus of the largest Lyapunov exponent included the values obtained in the study by Harbourne and Stergiou (2009). In this sense, values close to or less than zero (0) characterize a periodic signal/system with high periodicity and regularity. On the other hand, values close to 0.1 qualify chaotic signals/systems with high variability and complexity, where values ≥ 0.4 characterize a system with low regularity and high variability.

Although being evaluated in the context of golf putting, by applying the proposed methodology one can characterize any type of human movement in terms of regularity and stability. As such, this methodology can be used to assess the performance of an individual, by comparing it with the typical expected outcome provided by the approximate entropy and Lyapunov exponents. Moreover, as these measures allow the chaos of a given human process to be classified, it may shed some light into the closer relationship between process and product variables (Dias et al. 2014b).

5.2.4 Evaluation

This section presents the applicability of the previously presented non-linear methods. In that sense, Fig. 5.4 depicts the average approximate entropy for the motor execution of the putting of each player by considering the same experimental setup previously described. The average approximate entropy obtained by the 10 players in each study and respective distance of shot shows values that vary between 0.033 and 0.050. Players 1 and 9 proved to be the most consistent (present the lowest approximate entropy), whereas players 3, 6 and 8 presented the highest levels of entropy. In addition, when calculating the average of all the values of approximate entropy for each data set, the average value of approximate entropy for putting performance in expert players was 0.053. This is a very stable, regular and periodic value. Through the values obtained for the average of approximate

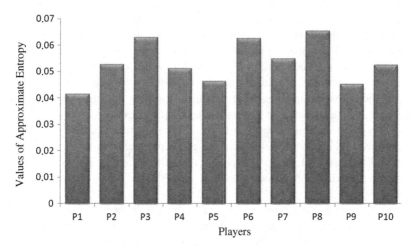

Fig. 5.4 Average of approximate entropy for motor execution of putting of each player in three studies (adapted from Dias et al. 2014b)

entropy, Fig. 5.4 shows a pattern of regularity and stability of players in the motor execution of putting throughout the three studies (Dias et al. 2014b).

Figure 5.5 presents the median of the Lyapunov exponent for the motor execution of the putting of each player in the three studies. The choice to analyze the data shown in Fig. 5.5 fell on the median, bearing in mind that the Lyapunov exponent can show extreme and negative values that influence the mean. Moreover, unlike the mean, which can disguise the results obtained, the median is a measure of central tendency that is more consistent and suitable to analyze the Lyapunov exponent. In other

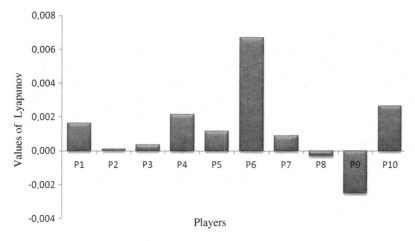

Fig. 5.5 Median of the Lyapunov exponent for motor execution of the putting of each player in three studies (adapted from Dias et al. 2014b)

words, considering the central value of data distribution, it was concluded that 50 % of the values are below the median and the other 50 % are above it (Stergiou et al. 2004; Dias et al. 2014b). The median of the Lyapunov exponent revealed values that were between −0.001 and 0.003. Player 9 presented a lower Lyapunov exponent, whilst player 6 reached the highest value. Moreover, considering the median of all the values of the Lyapunov exponent for each set of data, the resulting value from this non-linear tool for putting performance in players was 0.001 (Fig. 5.5).

As with approximate entropy it was possible to identify a pattern of regularity and stability of the players throughout the entire research. Player 9 presented the lowest and most stable Lyapunov exponent throughout the nine practice conditions in the three studies. Moreover, player 8 also showed negative values of the Lyapunov exponent.

By tuning into a non-linear approach and crossing the border into dynamic and chaotic systems, one may confirm that the players adapted to the variability and 'noise' that emerged from putting execution, and self-organized their performance towards the goal of the task (Davids et al. 2008). In this sense, the variability that results from motor performance can constitute a 'digital fingerprint' or 'putting signature' that is exclusive to each golfer (Couceiro et al. 2013; Dias et al. 2013). Nevertheless, this sort of analysis requires proficiency in multiple fields, from which engineering and mathematical skills are vital not only to adequately apply these (and other) non-linear methods, but to understand the output they provide. The following section presents a methodology which benefits from mathematical modeling, optimization and classification methods to provide a structured and semi-autonomous way of evaluating time-variant process variables.

5.3 Time Series Linear Analysis

As well as in the previous section, the intention here is to understand process variables as a whole instead of focusing on specific measures and properties of the time series (e.g., amplitude, duration, and others described in Sect. 5.1). Nevertheless, as opposed to non-linear methods, we present here a linear-based approach that comprises three subsequent phases: (i) mathematical modeling; (ii) optimization; and (iii) classification. Note that we will not fully describe, in any way, the literature around each of these subjects. We will thus provide the necessary concepts to use specific methods and tools within these subjects to analyze golf putting, mainly focusing on the horizontal trajectory of the putter.

5.3.1 Mathematical Modeling

Mathematically modeling a given phenomenon may be considered as the ultimate way to fully understand it—from plant growth (Prusinkiewicz 2004) to the flapping motion of a bird's wings (Couceiro et al. 2012a, b), there is the belief that everything around us can be represented in a mathematical manner (Gershenfeld 2011). Although this philosophy was extended to the general population in 1999 with the beginning of *The Matrix* movie triology, the concept of mathematical modeling is as old as mathematics itself. One of the few attempts to model a considerably complex phenomenon was carried out by Lorenz, with the help of physicists Howard and Markus, who formulated the mathematics behind the rotation of a water mill. Through experimentation, Lorenz was able to conclude that the mill is very susceptible to initial conditions and, therefore, he classified it as a chaotic system. By plotting the parameters of the mathematical model he then arrived to the well-known *butterfly effect* (Gleick 1987).

This was an important step with regard to mathematical modeling as well as to any mathematics applied to the study of the human movement. As previously stated, human movement is inherently a biological process and, as such, is hardly periodic (Stergiou and Decker 2011). Hence, one may consider that the case study at hand, golf putting, may be one of the few exceptions with a high periodic tendency. This was clearly seen in Figs. 4.10 and 4.11 (Sect. 5.2.1), in which the attractor shows how the putting dynamics fit between periodic and chaotic.

In this section we focus on presenting a mathematical model that may completely represent the horizontal trajectory of putting. Note that other process variables around golf putting (e.g., vertical trajectory, linear acceleration, etc.) could also be represented by the exact same mathematical model. However, for the sake of simplicity, and considering the previous analysis that was carried out around the horizontal trajectory of putting, we will focus on that one.

Nevertheless, by analyzing the shape of several trials from Fig. 4.10, it is clear that to model the putter's horizontal position in time, one ought to use a sinusoidal-like function. Nevertheless, a function composed by only one sinusoid was not precise enough to describe the movement, as it is clear in f_1 of Fig. 5.2, which results, in this case, in a MSE of 2.6568 units. This happens because the amplitude, angular frequency and phase of the descending half-wave, which corresponds to the player's backswing and downswing is usually different than the ascending half-wave, which corresponds to the ball's impact and follow-through. These disparities could not be represented using only one sinusoid wave. Therefore, to obtain a more precise model, a sum of sinusoid waves was employed. However, a compromise between precision and complexity of the problem had to be assumed, because each sinusoid adds three more dimensions to the estimation problem (amplitude, angular frequency and phase of the corresponding sine wave). In order not to let the complexity of the problem grow inappropriately, a function composed by the sum of three sinusoids was used (f_3 of Fig. 5.6), due to its precision, with a MSE of 0. 6926, when compared to using only a sum of two sinusoids, with a MSE of 0.7124 (f_2 of Fig. 5.6).

Thus, having the estimation function defined as a sum of three sine waves, each of the three parameters of each wave needs to be estimated, resulting in a nine dimension estimation problem which attempts to minimize the mean squared estimation error for every experiment, in order to obtain an accurate mathematical

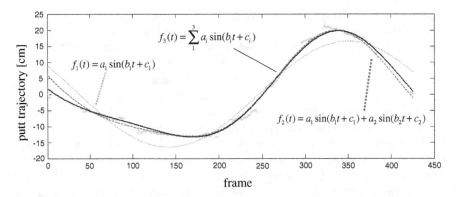

Fig. 5.6 Fitting sinusoidal functions to a point cloud, representing the position of a golf club during putting execution (adapted from Couceiro et al. 2013)

function that describes the horizontal position of the golf club during putting execution (Couceiro et al. 2013).

5.3.2 Optimization

Any mathematical model contains some parameters that can be used to fit the model to the real system it is intended to describe. This is commonly known as curve fitting, although a more general, and adequate designation would be *optimization*. Briefly, optimization is the selection of a best element (with regard to some criteria) from some set of available alternatives. When applied to fitting a given mathematical model to a real system, optimization is the process of finding the optimal parameter of such a model that has the best fit to a series of data points, possibly subject to constraints (Gill et al. 1981; Couceiro et al. 2013).

The search for an algorithm capable of dealing with most optimization problems without being very time-consuming and computationally demanding has been the subject of research in several scientific areas, such as control engineering and applied mathematics. Given the wide applicability of optimization besides fitting, the literature contains hundreds of methods, both traditional and non-traditional, in which biologically inspired methods fit within the latter. Here, we will briefly describe a biologically inspired method to find the most adequate model parameters to fit the horizontal trajectory of putting. This method was previously proposed by (Couceiro et al. 2012a, b) and denoted as *Fractional Order Darwinian Particle Swarm Optimization* (FODPSO). For more details about FODPSO, please refer to (Couceiro et al. 2012a, b) and (Ghamisi et al. 2014).

As in traditional PSO presented by Eberhart and Kennedy (1995), FODPSO candidate solutions (e.g., birds) are called particles. These particles travel through the search space to find an optimal solution, by interacting and sharing information with neighboring particles, namely their individual best solution (local best) and computing the neighborhood best. Furthermore, in each step of the procedure, the global best solution obtained in the entire swarm is updated. Using all of this information, particles realize the locations of the search space where success was obtained, and are guided by these successes. In each step of the algorithm, a fitness function is used to evaluate the particle success. For the problem at hand, i.e., find the parameters of a mathematical model to fit some series of data points, the particle success is defined by the one 'located' in the best configuration of parameters that result in the best fit to such data points. That particle is often known as global best particle and represents the optimal solution of the problem (when it exists).

Contrary to the PSO, the FODPSO comprises several swarms that compete between themselves, following Darwin's natural selection mechanism presented by Tillett et al. (2005). To model a given swarm, each particle n moves in a multi-dimensional space according to position $(x_n^s[t])$ and velocity $(v_n^s[t])$ values which are highly dependent on local best $(\chi_{1n}^s[t])$, also known as cognitive component, and global best $(\chi_{2n}^s[t])$, typically known as social component, as it follows:

$$v_n^s[t+1] = w_n^s[t] + \sum_{i=1}^{2} \rho_i r_i (\chi_{in}^s[t] - x_n^s[t]), \tag{5.13}$$

$$x_n^s[t+1] = x_n^s[t] + v_n^s[t+1] \tag{5.14}$$

$$w_n^s[t+1] = -\sum_{k=1}^{r} \frac{(-1)^k \Gamma[\alpha+1] v_n^s[t+1-kT]}{\Gamma[k+1]\Gamma[\alpha-k+1]}, \tag{5.15}$$

where Γ is the gamma function represented by $\Gamma(k) = (k-1)!$. Coefficients ρ_1 and ρ_2 assign weights to the local best and the global best when determining the new velocity, $v_n^s[t+1]$, respectively. Typically, the inertial influence is set to a value slightly <1. ρ_1 and ρ_2 are constant integer values, which represent *cognitive* and *social* components. However, different results can be obtained by assigning different influences for each component. For example, some works additionally consider a component called neighborhood best (which in this case would be ρ_3). The parameters r_1 and r_2 are random vectors with each component generally a uniform random number between 0 and 1. The intent is to multiply a new random component per velocity dimension, rather than multiplying the same component with each particle's velocity dimension. Equation (5.15) is built based on the concept of fractional calculus. Therefore, the order of the velocity derivative can be generalized to a real number $0 < \alpha < 1$, thus leading to a smoother variation and a longer memory effect when compared to other swarm intelligent approaches, such as traditional PSO (Couceiro et al. 2013).

Finally, we have an evolutionary metaheuristic computational method with multiple *coopetitive* particles, striving to find the optimal solution that may fit some data points, which is reflected upon a set of parameters evolving over time. For the specific case of golf putting, and considering the mathematical model chosen (see Fig. 5.2):

$$f(t) = a_1\sin(b_1 t + c_1) + a_2\sin(b_2 t + c_2) + a_3\sin(b_3 t + c_3), \qquad (5.16)$$

where the 'position' of each particle is represented by nine parameters (a_i, b_i, c_i), $i = \{1, 2, 3\}$ which yield a given 'solution'. Such a solution can be represented by the MSE, as previously considered, or any other measure of error, such as the sum of squared errors. For instance, considering the MSE as a measure of performance, the particle presenting the smaller MSE will be considered as the best performing particle of the swarm (Couceiro et al. 2013).

After a given number of iterations, or any other sort of stopping criteria (e.g., MSE minimum threshold), the method chooses the overall best particle from all best particles within each swarm. This results in a single mathematical function. Now comes the question: *what next?*

As opposed to Sect. 5.2, in which non-linear methods were applied to discrete observed data to further understand the putting execution, the fitting of the mathematical modeling outputs a single function from which one may benefit. Comparing two athletes can be as easy as simply comparing the nine parameters of each other's function. However, when facing multiple athletes within the chosen sample, such a comparison becomes more difficult. Moreover, although there are 'only' nine parameters for golf putting, other more chaotic actions (e.g., ballistic movements) may need a larger number of parameters inherent in more complex mathematical models. Although the dimension is only one order above the golf putting trajectory, the complexity of the problem grows substantially (one more dimension for each particle). Nevertheless, one should benefit from a method that has the potential to be applied to any function, regardless of the number of parameters it may possess, and regardless of the number of players one may wish to evaluate and compare. One way of achieving this is to use *classification* methods.

5.3.3 Classification

The skill in performing a putt is severely constrained by each person's profile and characteristics (Jonassen and Grabowski 1993). Furthermore, like other motor skills, individual performance of the putt results in a unique pattern (i.e., signature) that characterizes each player (e.g., Araújo et al. 2006; Schöllhorn et al. 2008). In order to better understand the concepts of 'signature' or 'fingerprint' of players in putting performance, as well as the differences within and between individuals, it is important to resort to computational methods of pattern recognition, also known as classification (Bishop 2006).

In this section, we merely present one of the many classification methods one could use—support vector machines (SVM). The choice of the method presented here resides on two points—the success of the method in the literature and its full implementation as an easy-to-use *MatLab Toolbox*[1] (Suykens et al. 2009).

SVM is a powerful technique for solving non-linear classification problems, function estimation and density estimation which has also led to many developments in Kernel-based methods. This method solves convex optimization problems, typically by quadratic programing. The LS-SVM is a reformulation to the standard SVM which was recently proposed. In fact, when the data points are linearly independent, the LS-SVM is equivalent to hard marginal SVM. LS-SVM involves the equality constraints only. Hence, the solution is obtained by solving a system of linear equations (Couceiro et al. 2013).

SVM models are similar to multilayer perceptron neural networks. However, using a Kernel function, SVMs are an alternative training method for polynomial, radial basis function (RBF) and multilayer perceptron classifiers, in which the weights of the network are found by solving a quadratic programing problem with linear constraints, rather than by solving a non-convex, unconstrained minimization problem as in standard neural network training. Furthermore, rather than fitting non-linear curves to the data, SVM handles this by using the kernel function to map the data into a different space where a hyperplane can be used to do the separation (Couceiro et al. 2013).

Many kernel mapping functions can be used but only a few have been found to work well in a wide variety of applications. The RBF is one of the most used and recommended kernel mapping functions in human movement studies. According to Hastie et al. (2009), kernel methods achieve flexibility by fitting simple models in a region local to the target point x_0. Localization is achieved via a weighting kernel K_λ, and individual observations receive weights $K_\lambda(x_0; x_i)$. RBFs combine these ideas, by treating the kernel functions $K_\lambda(\mu; x)$ as base functions, where each basis element is indexed by a location and a scale parameter (μ_m and λ_m, respectively). Thus, RBFs are symmetric p-dimensional kernels located at particular centroids:

$$f_\theta(x) = \sum_{m=1}^{M} K_{\lambda_m}(\mu_m, x)\theta_m. \qquad (5.17)$$

The centroids μ_m and scales λ_m have to be determined. A usual choice for the probability density functions is the standard Gaussian density function. There are also several approaches for learning the parameters μ_m, λ_m and θ_m. For example, a popular method is estimating θ_m, given μ_m and λ_m by a simple LS problem. Often the kernel parameters μ_m and λ_m are chosen in an unsupervised way using the X distribution alone. One of the methods is to fit a Gaussian mixture density model to the training x_i, which provides both the centers μ_m and the scales λ_m.

As with any other supervised classification method, there is a preliminary phase, called training, where one should create the training data, consisting of an input object (e.g., optimal parameters of the mathematical model) and a desired output

---

[1] http://www.esat.kuleuven.be/sista/lssvmlab/.

value (e.g., identification of the athlete). The SVM then analyzes the training data and produces an inferred function, which can be used for mapping new examples.

For instance, one way to study the variability of athletes while performing golf putting would be to use the "*all-but-one*" strategy, in which a random trial (i.e., set of nine parameters) is removed and tested by considering all remaining trials from that same athlete. The dimension of the region of the parameter will define the variability of a given athlete.

5.3.4 Evaluation

As a first step, let us evaluate the FODPSO method over four different optimization techniques: (i) the gradiant descent; (ii) the pattern search; (iii) the simplex; and (iv) the traditional PSO.

One popular first-order optimization algorithm used in many engineering-related works is the gradient descent. In this method, the search is carried through proportional steps in the direction of the negative of the gradient, or the approximate gradient, of the function at the current point to find a local minimum. It has been applied in the literature, for example, for face alignment in computer recognition systems. Another frequently used method is pattern search, which is a similar approach to the gradient descent; however, it does not compute the gradient, meaning that it can be used with non-differentiable functions. In this case, a descent search direction is produced by varying the parameters of the problem with different step sizes, aiming to obtain a fit to the experimental data. It has been successfully applied in model selection of SVMs and for transformation function search in automatic image registration etc. Both the gradient descent and the pattern search methods are more suitable for low-dimensional optimization problems (Couceiro et al. 2012a, b). In addition, the downhill simplex method was also considered a benchmark. This is an optimization method to numerically solve linear programing problems by searching for optimal solutions in the vertices of the admissible region of the space, considering all constraints, and iteratively improving the objective function. It is perhaps the most popular optimization algorithm for linear problems with low dimensions, being applied previously in contexts like particle accelerator control (Couceiro et al. 2012a, b). The traditional PSO was considered mainly to define a minimum performance threshold that the FODPSO should be able to overcome (Couceiro et al. 2013).

Three distinct trials of three different subjects were used in this first study. The goal was to check and compare the performance of each of the optimization techniques not mathematically representing the putting trials. All algorithms were run under the same conditions, with the same restrictions and with random initialization of all nine parameters. Figures 5.7, 5.8 and 5.9 and Table 5.3 clearly show the superior performance obtained by the FODPSO algorithm for all three cases, followed by PSO, which was only slightly inferior in the second dataset, when compared to pattern search and achieved the second best results in the first and third dataset. The other three algorithms generally obtained inferior results.

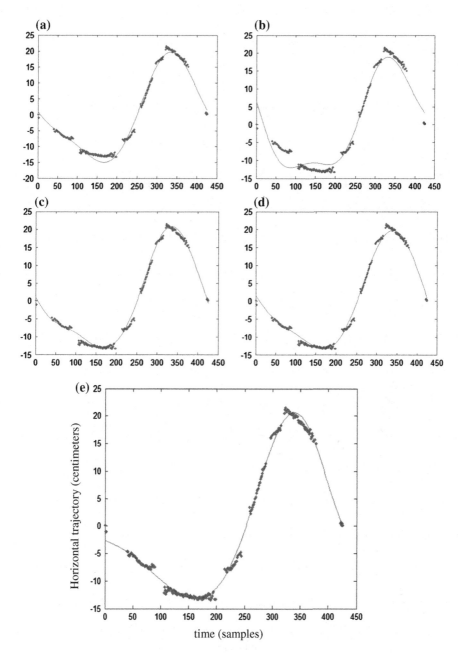

Fig. 5.7 Results for the first dataset. **a** Gradient descent; **b** pattern search; **c** downhill simplex; **d** PSO; **e** FODPSO (adapted from Couceiro et al. 2013)

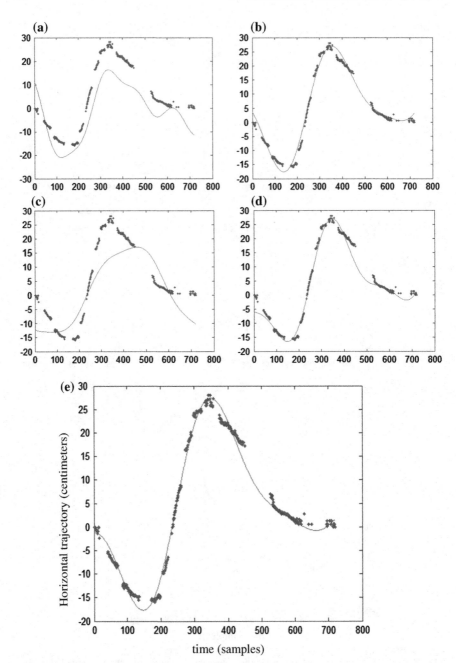

Fig. 5.8 Results for the second dataset. **a** Gradient descent; **b** pattern search; **c** downhill simplex; **d** PSO; **e** DPSO (adapted from Couceiro et al. 2013)

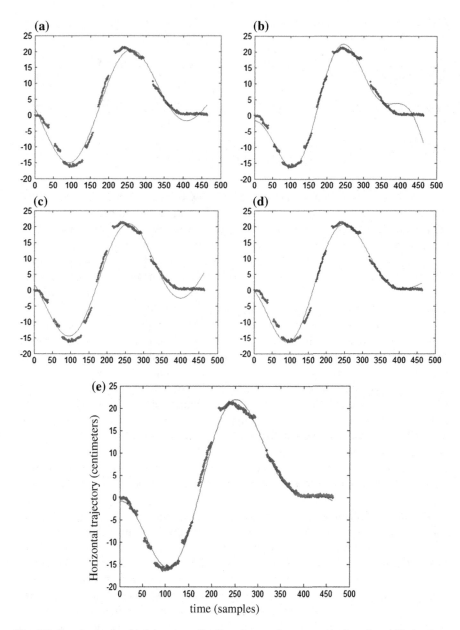

Fig. 5.9 Results for the third dataset. **a** Gradient descent; **b** pattern search; **c** downhill simplex; **d** PSO; **e** DPSO (adapted from Couceiro et al. 2013)

Table 5.3 Comparative results for all three datasets (Couceiro et al. 2012a, b)

	First dataset		Second dataset		Third dataset	
	Time (s)	*MSE*	Time (s)	*MSE*	Time (s)	*MSE*
Gradient descent	1,000	0.9270	1,000	7.2550	1,000	1.4989
Pattern search	1,000	1.7989	1,000	1.4315	1,000	1.4087
Downhill simplex	952	0.7147	1,000	5.7457	1,000	1.6770
PSO	312	0.6926	1,000	1.4391	1,000	0.9789
FODPSO	102	0.5658	1,000	1.0001	8	0.7368

Gray areas indicate better performance (adapted from Couceiro et al. 2012a, b)

Similar to the optimization phase, we will now present a benchmark around the classification methods tested within the context of golf putting. For improved comparison purposes, six athletes were considered. We will then compare the LS-SVM method with four other alternatives: (i) linear and quadratic discriminant analysis (LDA); (ii) quadratic discriminant analysis (QDA); (iii) NB with normal (Gaussian) distribution (NV); and (iv) NB with kernel smoothing density estimate (NVK) (Couceiro et al. 2013). The AUC was considered to evaluate the performance of the classification algorithms (Couceiro et al. 2013).

LDA is one of the most commonly used techniques for data classification and dimensionality reduction in statistics, pattern recognition and machine learning. This technique easily handles the case where the within-class frequencies are unequal and its performances have been examined on randomly generated test data. LDA maximizes the ratio of between-class variance to the within-class variance in any particular dataset thereby guaranteeing maximal separability (Couceiro et al. 2013). The resulting classifier implies that the decision boundary between pairs of classes is linear and a hyperplane when using more than two classes. Although they differ in their derivation, QDA is similar to LDA. The essential difference between them is in the way the linear function is fit to the training data. Another popular feature, QDA separates measurements of classes of objects or events with a boundary between each pair of classes described by a quadratic equation. NV is one of the most efficient and effective inductive learning algorithms for machine learning and data mining. The NV classifier is designed for use when features are independent of each another within each class, which is a rather unrealistic assumption that is almost always violated in real-life applications. However, it has a surprisingly good performance; performing well in practice even when the independence assumption is not valid. Furthermore, this assumption dramatically simplifies the estimation. The individual class-conditional marginal densities can be estimated separately; if the variables are discrete, then an appropriate histogram estimate can also be used. Bayesian classification and decision making are based on probabilities that a given set of measurements come from objects belonging to a certain class (probability theory). Statistical methods based on class conditional probability density functions of features, are suitable in diverse classification tasks. Estimated probability density functions have been used for classification utilizing Bayes' formula. The classification can be performed based on the probability

Table 5.4 Average value of the AUC for the five classification methods

Athlete	LDA	QDA	NV	NVK	SVM
1	0.619	0.601	0.671	0.680	0.744
2	0.650	0.623	0.692	0.685	0.737
3	0.566	0.582	0.634	0.761	0.734
4	0.507	0.585	0.574	0.675	0.690
5	0.622	0.651	0.692	0.766	0.797
6	0.493	0.602	0.650	0.718	0.745

Gray areas indicate better performance (adapted from Couceiro et al. 2012a, b)

density function, instead of estimating posterior probability using NV. The attempt is to estimate the underlying density function from the training data, and the idea is that the more data in a region, the larger is the density function. NVK leads naturally to a simple family of procedures for nonparametric density estimates for classification in a straightforward fashion using Bayes' theorem (Couceiro et al. 2013).

Table 5.4 compares the performance of the classifiers used to identify the signature of each player. According to the analysis of the AUC, the LS-SVM presented an overall better performance.

Based on these results, let us now extract the individual putting signatures using the LS-SVM method.

In Table 5.5, each color corresponds to a player (class) following the colors specified by the built-in *MatLab Colormaps 'lines'*. The colored areas correspond to the relationship between the parameters of the mathematical model previously defined in Eq. 5.4 for the 30 trials performed by each player.

As shown in Figs. 5.10, 5.11 and 5.12, each player's putting ability can be defined by six pairs of parameters related to the three sinusoids of the estimation model. Different players have different regions defined in the parameter's space, which in turn, indicate a difference in the game style of each player. However, as expected, intersections (misclassified data) between the regions in the parameter's space are common; this is due to the expertise of the players resulting in a similar AUC value of ~70 %. These results showed that classification algorithms could describe the motor performance of individual players, considering them as a unique 'signature' for each player. The scattering data also show that it was possible to categorize different ways of performing the golf putt. Although the data did not clarify if a player had a better or worse performance compared to others, it was clear that there are individual profiles that are unique to each player. In other words, a greater or smaller dispersion of data (colored area) may indicate whether a player

Table 5.5 Player-to-color correspondence (adapted from Couceiro et al. 2012a, b)

Player 1	Player 2	Player 3	Player 4	Player 5	Player 6

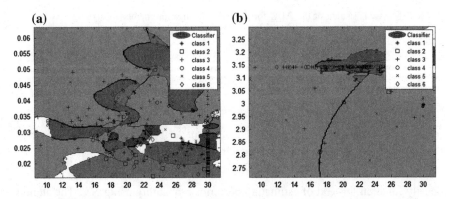

Fig. 5.10 Analysis of the first sine wave—*LS-SVM* classification of all six players. **a** Amplitude (a_1) versus angular frequency (b_1); **b** amplitude (a_1) versus phase (c_1) (adapted from Couceiro et al. 2012a, b)

was more consistent and regular than others during the motor execution. However, more research should be conducted in order to consolidate the results obtained in this work, and thus avoid falling into a classification error. For instance, player 6 showed a more consistent and unique putt action since it presented regions with a smaller area when compared to other players. This means that player 6 is more regular than the other players. The opposite is also true for player 4 and 5. The cyan and purple regions are scattered and usually form disconnected regions. As seen in the results, all players displayed unique putting characteristics. Similarities could be easily identified between them; however, the same could be stated about differences. These differences allowed the extraction of a signature for each player, confirming that a given putting stroke had a higher probability of belonging to a specific player.

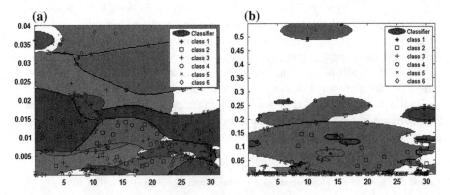

Fig. 5.11 Analysis of the second sine wave—*LS-SVM* classification of all six players. **a** Amplitude (a_2) versus angular frequency (b_2); **b** amplitude (a_2) versus phase (c_2) (adapted from Couceiro et al. 2012a, b)

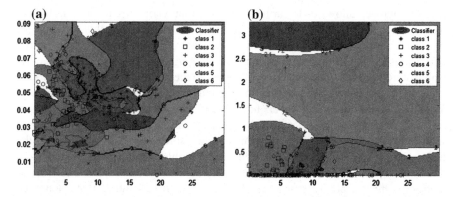

Fig. 5.12 Analysis of the third sine wave—*LS-SVM* classification of all six players. **a** Amplitude (a_3) versus angular frequency (b_3); **b** amplitude (a_3) versus phase (c_3) (adapted from Couceiro et al. 2012a, b)

5.4 Practical Implications

These findings point towards the need for new performance analysis methods around putting, where the individual analysis of kinematics is desirable over the traditional pooling of group data. In spite of this, the analysis of motor behavior profiles based on the measurement of individual kinematic strategies can help in better understanding the relevant changes resulting from the interaction between the athlete's characteristics and the surroundings where the task is performed.

Regarding the functional concept of variability and the contribution that the classification of 'signatures' may have in understanding the motor control of golf putting, it is considered important to further investigate the structure of the variability and influence that this movement may have in the performance of athletes with different skill levels.

Finally, it is certain that technological advances and new tools that recently emerged for human movement pattern analysis can make a strong contribution to the study of golf putting. Thus, we believe that this work may contribute to a deeper analysis of human motor behavior and performance in other sport movements.

5.5 Summary

This chapter was particularly important with regard to the study of movement and motor skills training, going beyond the analysis of the measures associated with the product and the response magnitude (e.g., position of the ball over the hole or quantification of the final result). We argue that it is important to address motor execution process variables when facing constraints (Newell 1986), and how the behavior of individuals changes and evolves over time (Davids et al. 2008). This type of approach allows us to further understand the variability inherent in human movement in the sports context. In our perspective, understanding this variability is crucial to the contemporary fields of motor control, sport sciences and biomechanics, especially in teaching, learning and training contexts. It is known that the way that golfers, and athletes in general, adapt to the constraints emerging from the task, is accomplished by adapting their motor execution velocity, acceleration, amplitude and duration. However, how this is really accomplished needs further research since, as we have seen in this chapter, this may cause multiple practice implications and decrease the performance of the athletes (e.g., final location of the ball to the hole). Therefore, it is considered highly important to benefit from advanced techniques (e.g., entropy, classification methods, and others) in order to demystify the variability structure inherent in motor execution that cannot be assessed using classical linear statistical methods.

References

Araújo D, Davids K, Hristovski R (2006) The ecological dynamics of decision making in sport. Psychol Sport Exerc 7(6):653–676. doi:10.1080/17461391.2014.928749

Bishop CM (2006) Pattern recognition and machine learning (information science and statistics). Springer, New York

Coello Y, Delay D, Nougier V, Orliaguet JP (2000) Temporal control of impact movement: the "time from departure" control hypothesis in golf putting. Int J Sport Psychol 31(1):24–46. ISSN 0047-0767

Couceiro MS, Dias G, Martins FML, Luz JM (2012a) A fractional calculus approach for the evaluation of the golf lip-out. SIViP 6(3):437–443. doi:10.1007/s11760-012-0317-1

Couceiro MS, Portugal D, Gonçalves N, Rocha R, Luz JM, Figueiredo CM, Dias G (2012b) A methodology for detection and estimation in the analysis of golf putting. Pattern Anal Appl 16:459–474. doi:10.1007/s10044-012-0276-8

Couceiro MS, Dias G, Mendes R, Araújo D (2013) Accuracy of pattern detection methods in the performance of golf putting. J Mot Behav 45(1):37–53. doi:10.1080/00222895.2012.740100

Davids K, Button C, Bennett SJ (2008) Dynamics of skill acquisition—a constraints-led approach. Human Kinetics Publishers, Champaign

Delay D, Nougier V, Orliaguet JP, Coello Y (1997) Movement control in golf putting. Hum Mov Sci 16(5):597–619. doi:10.1016/S0167-9457(97)00008-0

Dias G, Mendes R, Couceiro MS, Figueiredo C, Luz JMA (2013) On a ball's trajectory model for putting's evaluation, computational intelligence and decision making—trends and applications, from intelligent systems, control and automation: science and engineering bookseries. Springer, London

Dias G, Couceiro MS, Barreiros J, Clemente FM, Mendes R, Martins FM (2014a) Distance and slope constraints: adaptation and variability in golf putting. Mot Control 18(3):221–243. doi:10.1123/mc.2013-0055

Dias G, Couceiro M, Clemente F, Martins F, Mendes R (2014b) A new approach for the study of golf putting. S Afr J Res Sport Phys 36(2):61–77. ISBN 0379-9069

Eberhart R, Kennedy J (1995) A New Optimizer Using Particle Swarm Theory. Proc of 6th International Symposium on Micro Machine and Human Science, Nagoya, Japan. IEEE Service Center Piscataway NJ: 39–43

Gershenfeld N (2011) The nature of mathematical modeling. Cambridge University Press, Cambridge

Ghamisi P, Couceiro MS, Martins FM, Benediktsson JA (2014) Multilevel image segmentation based on fractional-order darwinian particle swarm optimization. IEEE Trans Geosci Remote Sens 52:2382–2394

Gill PE, Murray W, Wright MH (1981) Practical Optimization Academic Press

Gleick J (1987) Chaos: making a new science, Viking Penguin, New York

Grassberger P, Procaccia I (1983) Characterization of strange attractors. Phys Rev Lett 50(5):346–349. doi:10.1103/PhysRevLett.50.346

Harbourne RT, Stergiou N (2009) Movement variability and the use of nonlinear tools: principles to guide physical therapist practice. JNPT Am Phys Ther 89(3):267–282. doi:10.2522/ptj.20080130

Hastie T, Tibshirani R, Friedman J, Hastie T, Friedman J, Tibshirani R (2009) The elements of statistical learning, vol 2, no 1. Springer, New York. doi:10.1007/978-0-387-84858-7

Hume PA, Keogh J, Reid D (2005) The role of biomechanics in maximising distance and accuracy of golf shots. Sports Med 35(5):429–449. doi:10.2165/00007256-200535050-00005

Jonassen DH, Grabowski BL (1993) Handbook of individual differences, learning and instruction. Lawrence Erlbaum, Hillsdale

Karlsen J (2003) Golf putting: an analysis of elite-players technique and performance. Dissertation, Norway School of Sport Sciences

Karlsen J, Smith G, Nilsson J (2008) The stroke has only a minor influence on direction consistency in golf putting among elite players. J Sports Sci 26(3):243–250. doi:10.1080/02640410701530902

Newell KM (1986) Constraints on the development of coordination. In: Wade MG, Whiting HTA (eds) Motor development in children: aspects of coordination and control. Martinus Nijhoff, Boston

Pincus S, Singer BH (1998) A recipe for randomness. Proc Natl Acad Sci USA 95(18):10367–10372

Pincus S, Gladstone MIM, Ehrenkranz RA (1991) A regularity statistic for medical data analysis. J Clin Monit 7(4):335–345. doi:10.1007/BF01619355

Prusinkiewicz P (2004) Modeling plant growth and development. Curr Opin Plant Biol 7(1):79–83

Rosenstein MT, Collins JJ, Luca CJ (1993) A practical method for calculating largest Lyapunov exponents from small data sets. Phys Nonlinear Phenom 65(1–2):117–134. doi:10.1016/0167-2789(93)90009-P

Sato S, Sano M, Sawada Y (1987) Practical methods of measuring the generalized dimension and the largest Lyapunov exponent in high dimensional chaotic systems. Prog Theor Phys 77:1–5

Schöllhorn W, Mayer-Kress G, Newell KM, Michelbrink M (2008) Time scales of adaptive behavior and motor learning in the presence of stochastic perturbations. Hum Mov Sci 28 (3):319–333. doi:10.1016/j.humov.2008.10.005

Stergiou N, Decker LM (2011) Human movement variability, nonlinear dynamics, and pathology: is there a connection? Hum Mov Sci 30(5):869–888. doi:10.1016/j.humov.2011.06.002

Stergiou N, Buzzi UH, Kurz MJ, Heidel J (2004) Non-linear tools in human movement. In: Stergiou N (ed) Innovative analyses of human movement. Human Kinetics Publishers, Champaign, pp 163–186

Suykens JAK, Van Gestel T, De Brabanter J, De Moor B, Vandewalle J (2009) Least squares support vector machines. World Scientific, Singapore

Takens F (1981) Detecting strange attractors in turbulence. In dynamical systems and turbulence, Warwick 1980. Springer, Berlin, pp 366–381. doi:10.1007/BFb0091924

Tillett J, Rao TM, Sahin F, Rao R, Brockport S (2005) Darwinian particle swarm optimization. In: Prasad B (ed) Proceedings of the 2nd Indian international conference on artificial intelligence. Pune, India, pp 1474–1487

Chapter 6
Conclusions

Abstract Writing a book about golf putting can be as complicated as performing the action itself; it is only after the ball goes into the hole that one can truly know that he/she has succeeded. However, one thing is certain, this short book presents multiple contributions to other researchers, players and coaches who may wish to investigate the variables behind the motor execution of this movement, which is apparently much more complex than one might expect. This final chapter summarizes the main contributions. It also delineates the limitations inherent in the several studies presented in the book, suggesting future directions to readers or researchers. The chapter ends by presenting the implications of these studies on research, education and training contexts.

Keywords Golf putting · Contributions · Discussion · Future work · Implications

In simple terms, golf putting can be learned in many different ways and with many different methods. Nevertheless, the individual variability underlying human motor behavior makes this movement quite different from player to player, since the morphological and functional characteristics are distinct. We verify that expert golfers can learn to putt under several conditions of practice variability. However, it should be noted that the golf putting performance substantially varies from player to player, depending on their morphological and functional features. In that sense, we concluded that morphological (height, weight and height) and functional (motivation, fatigue, etc.) characteristics of the players may influence the angular displacement applied to putting and, as a consequence, the force, the acceleration and the velocity of the movement throughout the putting performance. These aspects relate to the characteristics and profiles that distinguish each individual practitioner in the motor execution process.

On the other hand, we concluded that the irregularities from the green (slopes and texture of the grass) may also compel the golfer to adjust his/her technique to overcome the restrictions imposed on the task, thus inevitably influencing the putting performance. All these factors have a decisive influence on how the golfer will hit the ball and adjust its action, namely, the angular displacement of the putter

© The Author(s) 2015

G. Dias and M.S. Couceiro, *The Science of Golf Putting*,
SpringerBriefs in Applied Sciences and Technology,
DOI 10.1007/978-3-319-14880-9_6

and all other related process variables of motor execution. Moreover, golf putting also encompasses other relevant variables within the performance context, such as stability, routine, attitude and rhythm, as well as other aspects of personality, learning ability and motivation in the execution of this movement. One way to promote this is to film the motor performance and analyze it later to identify and correct any technical inconsistencies.

It is important to promote a gradual increase of the contextual interference in both learning and evaluation of golf putting. Under this premise, we can conclude that it is necessary to create an optimal learning/adapting level, such that an inexperienced golfer can adapt to the complexity of golf putting. In that sense, the characteristics of the task and the level of expertise of the athlete are both variables that one needs to consider during the design of the practice and motor learning in the context of golf putting. Finally, the several metrics presented in this book bring implications to the area of sports training since they aim at providing a deeper understanding of a player's flaws. These approaches are important mainly in a coaching perspective to avoid overusing standard metrics that lack relevant information about a given action. Although most of the traditional research around sport science is centered on the product variables, many researchers have been working toward a better understanding of the process measurements of motor execution.

6.1 Main Contributions

In 2000, Dave Pelz, with his *'Putting Bible'*, paved the way for other authors to deepen the psychological and physiological aspects behind golf putting. In this book, we have followed the accuracy and incisiveness of Pelz, sharing a natural extension of his bible with others who are interested in studying this motion in the laboratory or real-game context, for research, teaching and learning. Undeniably, every time one analyzes a 'small' process or product variable associated with putting, a series of questions and doubts appear. It is only natural that the motor performance and biomechanical aspects of the action have much to offer in terms of what a golfer can or cannot do, in both training and competition, and depending on his/her perception of the environment. Such complexities cannot be really assessed by merely considering the final accuracy of the action (i.e., if the player failed or hit the hole). Instead, this book paves the way towards a description of the reasons 'behind' this failure/success. In other words, we have shown that there is an emergent need to comprehend more than only the most standard product variables, thus attempting to determine to what extent the process variables (e.g., speed, acceleration and position) can ultimately influence the final outcome. To do so, one needs to resort to new ever-improving technologies. Such technologies can provide important feedback to the coach about how the player performs the movement and how his/her motor performance evolves over time. With this increasing amount of information, which was previously relatively unknown, one can conclude that the variability of practice conditions and constraints of the task can 'tune' and 'calibrate' the performance of the player, making him/her more proficient and immune to the unpredictability offered by the green and other environmental factors. The several methodologies, studies and approaches presented in this short book should be considered as guidelines that can go way beyond golf putting.

6.2   Limitations and Further Research

As with any research, this book also has some limitations. One of its major limitations is the level of multidisciplinary requirement needed to fully understand the range of scientific concepts and methodological aspects that the book offers. Although it was not our aim to prepare a manual for 'scientists', but to all people who may be interested in golf putting, the truth is that sport science, mathematics, engineering and computer science are some of the disciplines which this book resorts to. Either way, it is our expectation that readers can improve their understanding around golf putting and see beyond the 'place-the-ball-in-the-hole' philosophy.

That being said, further research should be conducted to remain on top of the insights provided in this book, so as to optimize a golfer's performance under different conditions of practice, environmental and contextual variability. To that end, a refined analysis with advanced tools and professional software should be considered to evaluate all the biomechanical components of the movement, giving effective answers about the kinetic and kinematic chains behind golf putting.

Another important aspect is to realize the extent to which performance metrics presented in this book are likely to be used in other sports and how such transfers can be made, for instance, to ballistic movements (e.g., golf swing).

Finally, the development of new technologies to further understand and improve human performance is essential. Given the positive feedback from many golf players, namely from the European Champion of *pitch and putting* Hugo Espirito Santo, *InPutter* is one of those new internet-connected technologies being currently used not only within research context, but in real-life golf training. More *out-of-the-box* technologies without the need of expert knowledge or exhaustive setups should be developed.

6.3   Implications for Research, Education and Training

The practical implication of this book is essentially defined by the description of a scientific approach applied to golf putting analysis. However, the main target audience goes beyond sport scientists, being aimed at students, teachers, players and coaches who want to learn more about golf putting. We are aware that this kind of research and its understanding is difficult without the benefits of having a multidisciplinary team. Nevertheless, most concepts are required, not only for the analysis of the action itself, but to compare a player's performance. This book can also be seen as a 'guide' for those who have begun to take the first steps to starting their own research, regardless of whether or not it is applied in the context of golf putting. This book describes multiple experimental procedures, such as the description of some of the most common acquisition technologies available. Besides being a guide, it is also an important educational support for children and young players who want to start to learn golf putting, having at their disposal some illustrations and concepts. There is a lack of tutorial support that the authors intend to further exploit in another book, namely the pedagogical progressions, exercises and 'tips' that may be useful for those who want to refine their psychomotor performance in the context of training and competition. However, time is limited and, in the future, another attempt to put the ball into the hole will be made.

So wait for the hit...

Acknowledgments The authors would like to thank Professor Rui Mendes, Professor João Barreiros, Professor Keith Davids, Professor Guilherme Lage, Professor Hugo Espírito Santo, Professor Nuno Barreto, Eng. André Araújo, Eng. Samuel Pereira and Mr. António Dias for theirtechnical support.